RING BOOKS

PILOT PACK

Volume 3

Also available

Pilot Pack Volume 1
Great Yarmouth to Littlehampton and IJmuiden to Carentan
0 229 11798 8

Pilot Pack Volume 2
Chichester to Portland; The Channel Islands;
St Vaast to Erquy
0 229 11799 6

PILOT PACK

Volume 3

Bridport to the Scilly Isles
Le Légué to Ushant

EDITED BY
BRIAN GOULDER

ADLARD COLES
8 Grafton Street, London W1

Adlard Coles
William Collins Sons & Co. Ltd
8 Grafton Street, London W1X 3LA

First published in Great Britain by
Adlard Coles 1989

British Library Cataloguing in Publication Data

Pilot Pack.
Vol. 3, Bridport to the Scilly Isles and
Le Légué to Ushant.
1. Western Europe. Coastal waters – Pilots' guides – For yachting
I. Goulder, Brian
623.89'294'0247971

ISBN 0-229-11800-3

Printed and bound in Great Britain by
Butler & Tanner Ltd, Frome, Somerset

Contents

Introduction

This Adlard Coles *Pilot Pack* is intended to provide yachtsmen, fishermen and skippers of small vessels with a set of specially drawn harbour and anchorage plans based on the latest information so that, with only the addition of passage charts and an almanac, the navigator has on board every work of reference needed for safe voyaging.

Graphs are provided for quick and accurate tidal calculations.

Every prudent sailor knows that charts and other printed information may contain errors; shore lights change, channels silt up or are dredged, breakwaters are built or demolished and buoys laid or removed. The publishers offer Amendment Notes which are up-dated weekly and which therefore provide the very latest corrections and amendments; readers are urged to take advantage of this service. In this context, the publishers will always be grateful to readers who draw attention to errors or who make suggestions for improvements. Correspondents are sought from each area covered within the *Pilot Packs* so that the information contained is kept under permanent review. These correspondents will receive a complimentary copy of any volume of the *Pilot Packs* or a copy of any Adlard Coles book of equivalent value.

ACKNOWLEDGEMENTS

Navigational information is based on Admiralty charts, Crown Copyright, reproduced with the permission of the Controller of HM Stationery Office, the Hydrographer of the Navy, copyright reserved, and the Service Hydrographique et Océanographique de la Marine, France, reproduced by permission. Tidal predictions for Dover have been computed by the Proudman Oceanographic Laboratory, copyright reserved.

DISCLAIMER

The greatest possible care has been taken with the compilation of this *Pilot Pack* and all the information is, at the time of going to press, accurate to tbe best of the publishers' knowledge and belief. However, neither the publishers, nor their agents, can accept any responsibility for any errors, omissions or subsequent changes, nor for the consequences of any incident in which the users of the *Pilot Pack* may be involved.

Area Map

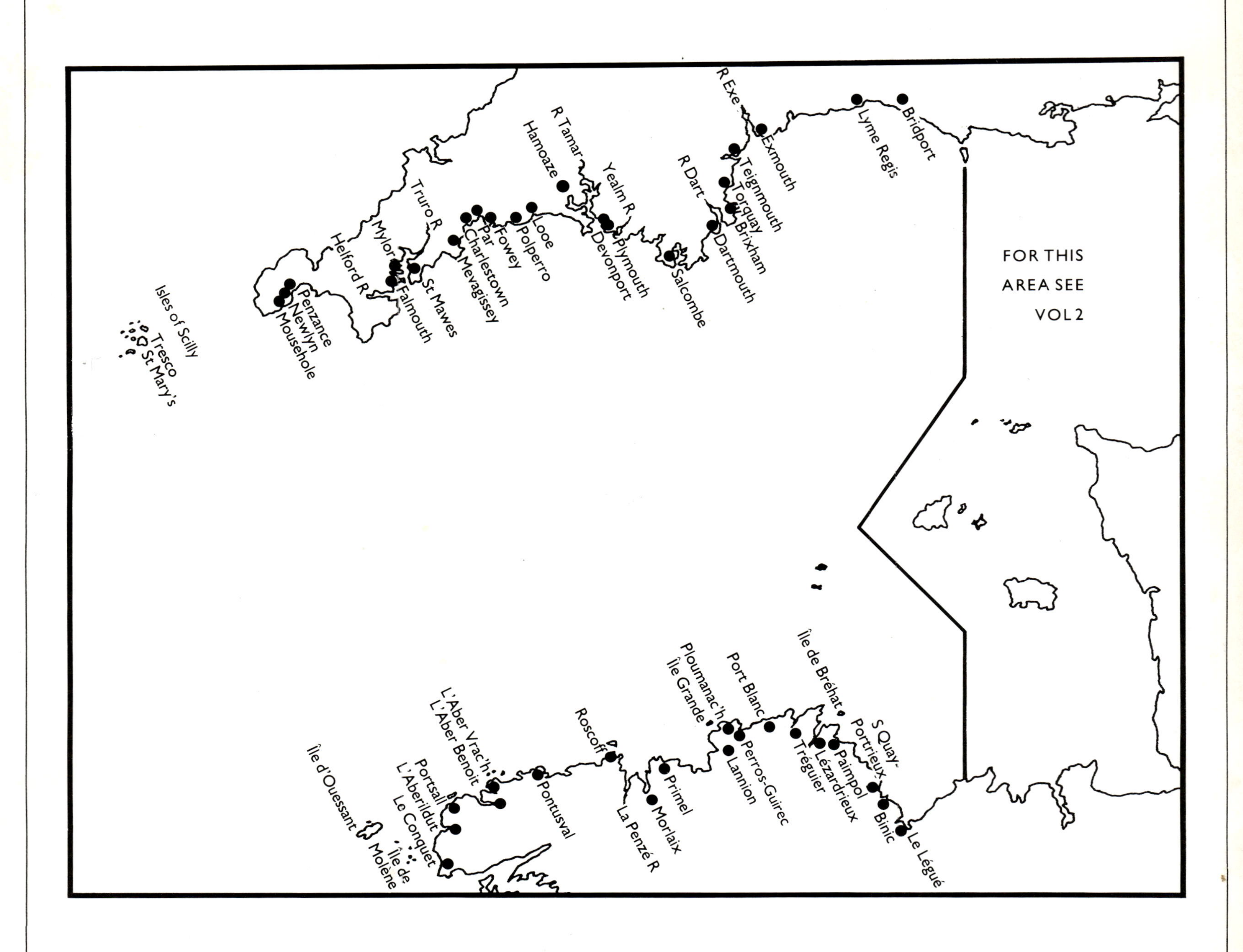
R Exe
Exmouth
Lyme Regis
Bridport
R Tamar
Hamoaze
Yealm R
Teignmouth
Torquay
Brixham
R Dart
Dartmouth
Salcombe
Plymouth
Devonport
Looe
Polperro
Fowey
Par
Charlestown
Mevagissey
Truro R
Mylor
Helford R
St Mawes
Falmouth
Penzance
Newlyn
Mousehole
Isles of Scilly
Tresco
St Mary's
FOR THIS AREA SEE VOL 2
Île de Bréhat
S Quay-Portrieux
Paimpol
Lézardrieux
Tréguier
Port Blanc
Perros-Guirec
Ploumanac'h
Île Grande
Lannion
Primel
Morlaix
La Penzé R
Roscoff
L'Aber Vrac'h
L'Aber Benoit
Pontusval
Portsall
L'Aberildut
Le Conquet
Île d'Ouessant
Île de Molène
Binic
Le Légué

Symbols and Abbreviations

PORTS AND HARBOURS

- Customs House
- Harbour Master's Office
- Yacht Marina
- Fuel
- Yacht Club Anchorage
- Water

BUILDINGS

	Church
	Monument
	Chimney
Tr	Tower
	Water Tower
CG	Coast Guard
	Windmill
SS	Signal Station
	Flagstaff

LIGHTS

	Light
	Minor light
F	Fixed
Oc	Occulting
Oc(2)	e.g. Group occulting (2)
Iso	Isophase
Fl	Flashing
LFl	Long-flashing
Fl(3)	e.g. Group flashing (3)
Q	Quick
Q(3)	e.g. Group quick (3)
VQ	Very quick
Alt	Alternating
hor	Horizontal
vert	Vertical
Ldg Lts	Leading Lights

COLOURS

W	White
R	Red
G	Green
Y	Yellow
Y or **Or**	Orange
B	Black
Bu	Blue
Vi	Violet

BUOYS AND BEACONS

- Can buoy
- Conical buoy
- Spherical buoy
- Spar buoy
- North cardinal buoy
- South cardinal buoy
- East cardinal buoy
- West cardinal buoy
- Mooring buoy

TOP MARKS

- Starboard hand
- Port hand
- Special
- Isolated danger
- Safe water

DANGERS AND PROHIBITIONS

(1₂)	Rock which covers and uncovers (with height above chart datum)
	Rock awash at chart datum
+	Rock with 2m or less over it at chart datum
	Wreck showing at chart datum
	Wreck with masts visible
7₃ Wk	Wreck with depth by sounding
9₁ Wk	Wreck with depth by sweep
	Submarine cable
	Submarine pipeline
	Anchoring prohibited
	Fishing prohibited

DEPTHS AND TYPE OF BOTTOM

	Dries less than 2m
----	5m line
....	10m line
S	Sand
M	Mud
Cy	Clay
G	Gravel
Sn	Shingle
R	Rocks
bk	Broken
Sh	Shells
St	Stones
HW	High Water
LW	Low Water
MHWS	Mean High Water Springs
MLWS	Mean Low Water Springs
MHWN	Mean High Water Neaps
MLWN	Mean Low Water Neaps
LAT	Lowest Astronomical Tide

ABBREVIATIONS ALSO USED IN THE TEXT

M	Nautical mile or Magnetic
m	metre
s	second
Lt	Light
MRCC	Maritime Rescue Coordination Centre
MRSC	Maritime Rescue Sub-Centre
YC	Yacht Club
SC	Sailing Club
Pt	Point
by	buoy
Ch	Church or Channel
occas	occasional
IALA	International Association of Lighthouse Authorities
Ro Ro	Roll-on Roll-off ferry
D/F	Direction Finding
Ru	Ruin
Br	Bridge
Chy	Chimney

Traffic Signals

IALA traffic signals are gradually coming into use and will, it is hoped, eventually become standard

LIGHTS

● Red flashing ● Red flashing ● Red flashing	Serious emergency. All vessels stop
● Red fixed or occulting ● Red fixed or occulting ● Red fixed or occulting	Vessels not to proceed
● Green fixed or occulting ● Green fixed or occulting ● Green fixed or occulting	Vessels may proceed; one way traffic
● Green fixed or occulting ○ White fixed or occulting	Vessels may proceed; two way traffic
● Green fixed or occulting ○ White fixed or occulting ● Green fixed or occulting	A vessel may proceed only if she has received specific orders to do so

COMMON FRENCH AND BELGIAN TRAFFIC SIGNALS

Lights *full*	Lights *simplified*	Flags or Shapes *full*	Flags or Shapes *simplified*	Meaning
● Red ○ White ● Red	● Red		R	Entry prohibited
● Green ○ White ● Red	● Red ● Green		R G	Entry and departure prohibited
● Green ○ White ● Green	● Green		G	Departure prohibited

Balls

Lights	Balls	Meaning
● Red ● Red ● Red	● Red ● Red ● Red	Emergency – entry prohibited
● Green ● Green ● Green		Port open

Notes on the use of the *Pilot Pack*

Each pilot page is set out in a uniform and self explanatory manner with the relevant chart on the facing page.

WAY POINTS

Way points suggested may not give safe passage from every direction and a track must always be laid off to ensure that there are no dangers.

VARIATION

Variation is rounded to the nearest half degree.

VHF

Call signals are ordinarily addressed to a station by its name, unless the call is shown in capital letters, in which case this is the call sign which should be used.

TIDAL DEFINITIONS

Definitions of the various states of level of the sea and heights of land features which relate to it are illustrated below.

CHART DATUM

Chart datum on *Pilot Pack* charts is Lowest Astronomical Tide (LAT), therefore there will almost always be more water than the chart shows.

SOUNDINGS

Soundings and all calculations and references to depth of water are in metres.

TIDAL TIME DIFFERENCE ON DOVER

The difference between local High Water and HW Dover is given on the pilotage information page for most ports. This is approximate and should be used only if local tide tables are not available.

Similarly the times and directions of the tidal streams are based on HW Dover and are intended for passage planning rather than accurate dead reckoning navigation.

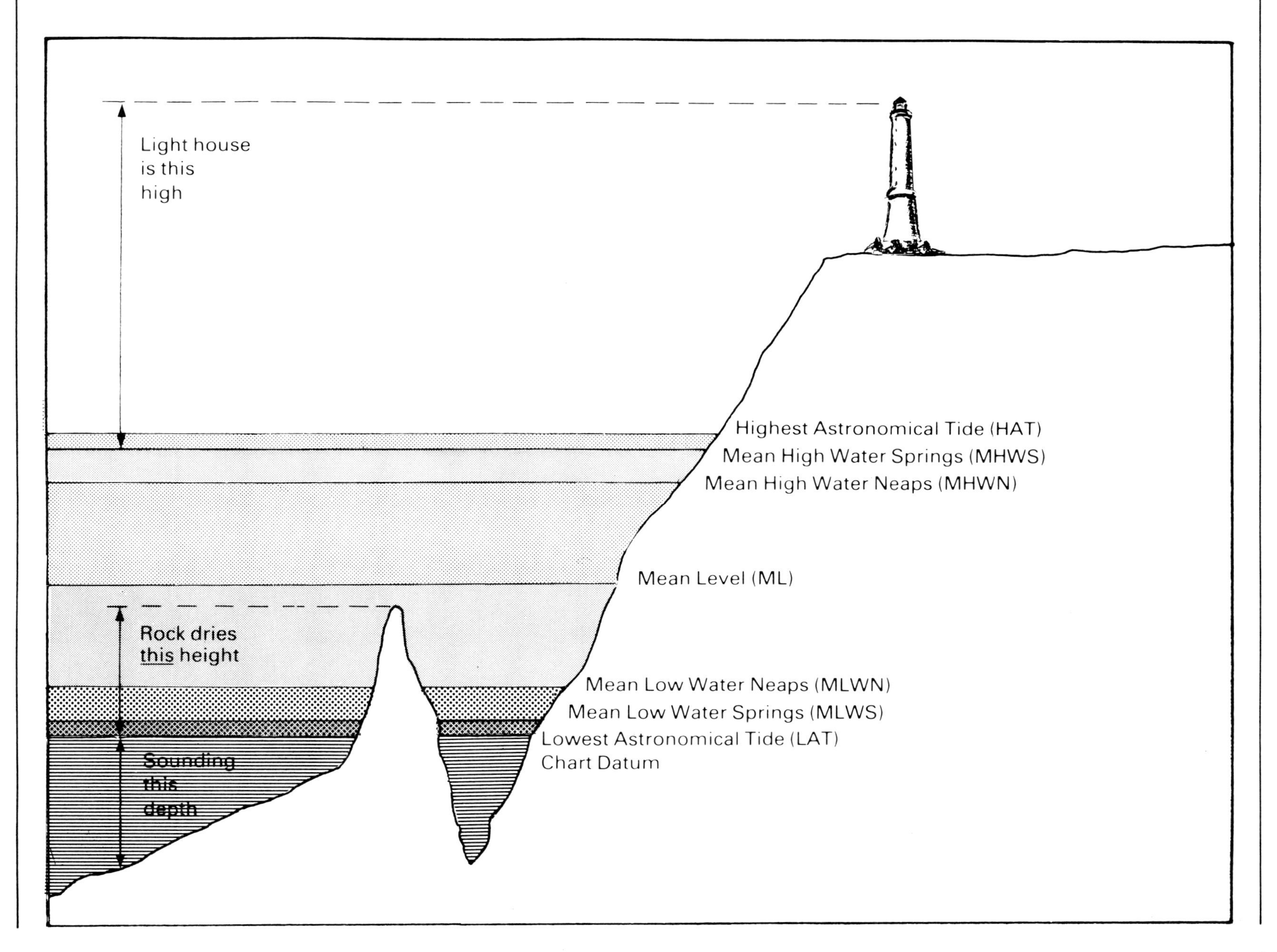

Notes on the use of the *Pilot Pack*

USE OF THE TIDAL CURVES

To determine height at a particular time, or time for a particular height.

This graphical method is much quicker and more accurate than previous methods and deserves the few minutes necessary to appreciate how simple it is to use.

The examples below assume that tide tables are available for the Standard Port. If not, use the Dover Tide Tables and correct for the local HW using the differences given on the pilot page.

Example 1: Standard Port
Find the height of tide at Plymouth (Devonport) at 1000 on Thursday 20 July 1989 and between what times the height of tide is 3.0m.

From the Plymouth (Devonport) tide tables:
HW 5.1m at 0658; LW 0.9m at 1257

Mark the time of HW in the box below the tidal curve and each successive hour before and after. Note that the range is (5.1 – 0.9) equals 4.2m, i.e. Springs. Mark the height of HW and LW on the upper and lower horizontal axes of the graph and draw a line between the two points. Rule upwards from 1000 (effectively 0958) to meet the full (Springs) curve, then horizontally to meet the diagonal line between the HW and LW and vertically to meet the top horizontal axis. This shows that there will be 3.4m (approx) above charted depth at 1000.

Rule down from the 3.0m point on the top horizontal axis to the diagonal and then across to meet both sides of the full (Springs) curve, thence down to the time scale. This shows that there will be 3.0m above charted depth between HW – 3hrs 40mins (i.e. 0320 approx) and HW + 3hrs 20mins (i.e. 1020 approx).

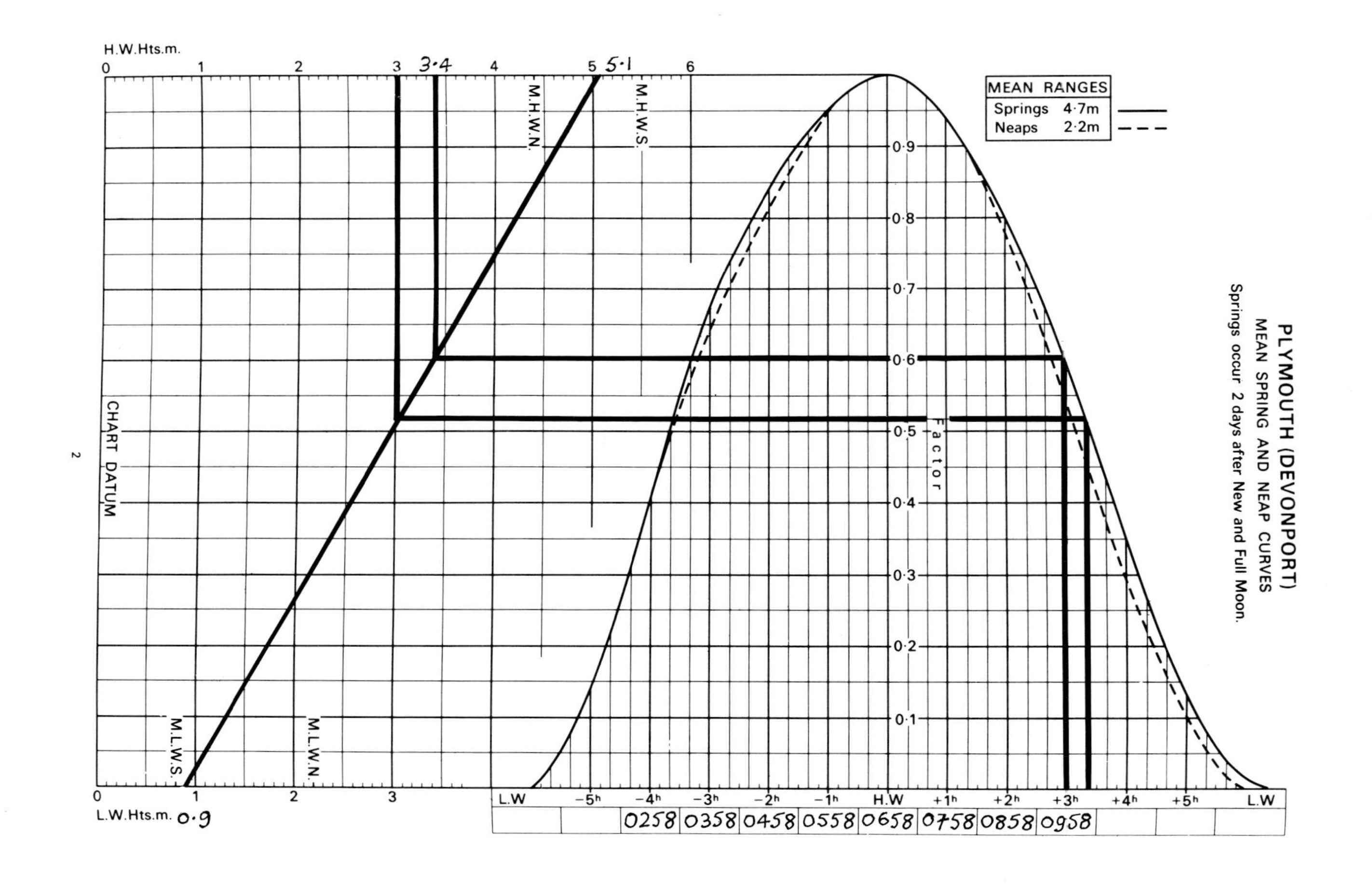

Example 2: Secondary Port
Find the height of tide at Newlyn at 1000 on Thursday 20 July 1989.

The Plymouth (Devonport) tidal data is as above. The differences for Newlyn are shown in the table included in the pilotage information, but for convenience reproduced below:

	Lat	Long	Time Difference HW		Time Difference LW		Height Difference (metres)			
PLYMOUTH See p. 28	50°22′N	4°11′W	0000 and 1200	0600 and 1800	0000 and 1200	0600 and 1800	5.5	4.4	2.2	0.8
Newlyn	50°06′N	5°33′W	−0040	−0105	−0045	−0020	+0.1	0.0	−0.2	0.0

TIDES relative to HW Dover (approx) HW +0600.

These differences are also shown graphically where only one or two secondary ports are referred to on a single pilotage page. In this case, Newlyn has completed charts showing the height and time differences.

Thus, when Plymouth (Devonport) has a HW height of 5.1m, there is a difference of +0.1m at Newlyn, i.e. the HW height at Newlyn will be 5.1 + 0.1 = 5.2m. There is no difference at LW which therefore remains at 0.9m. These two figures, 5.2m and 0.9m provide the diagonal.

When HW at Plymouth (Devonport) is 0658, the time difference at Newlyn is −0100, i.e. HW Newlyn on Thursday 20th July is 0658 − 0100 = 0558. This time is inserted into the box for HW and the other hours also inserted back to 0958. A vertical line from 0958 (near enough to 1000) to the full (Springs) curve, a horizontal line across to the diagonal and another vertical line up to the horizontal axis shows that at 1000 there will be 2.3m above charted depth at Newlyn.

This procedure can be reversed to give times for heights as above.

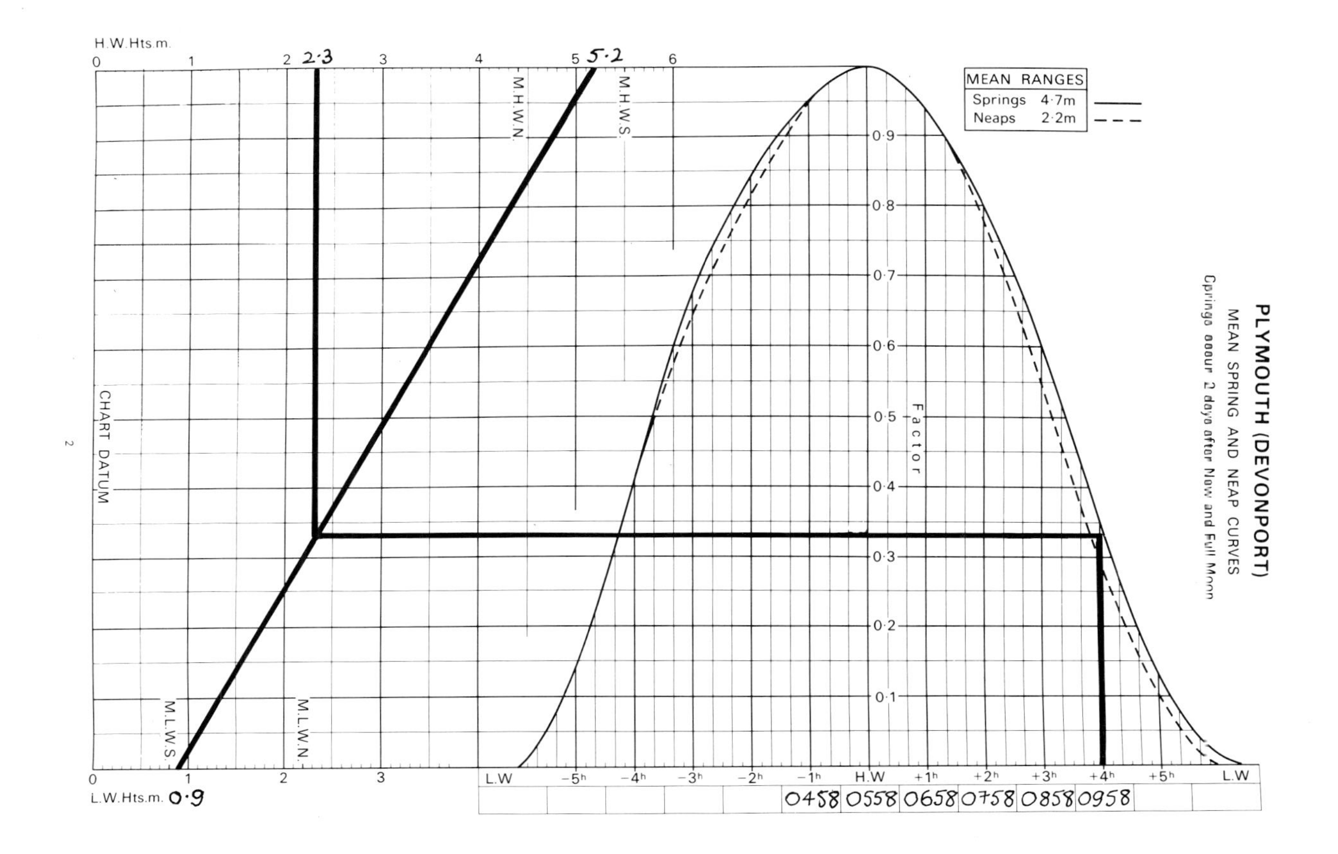

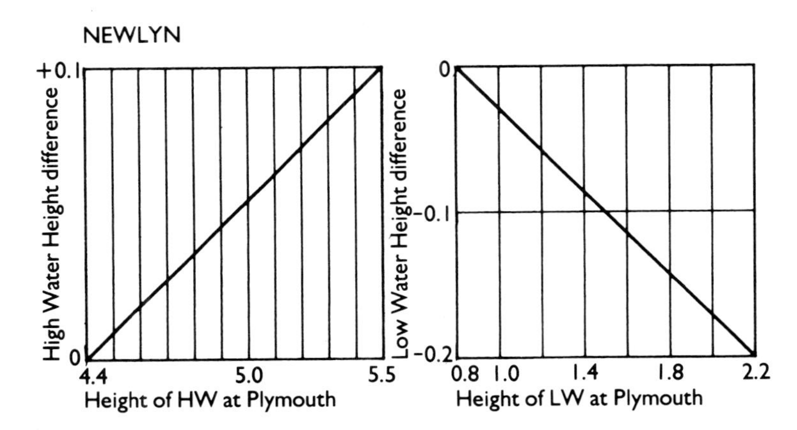

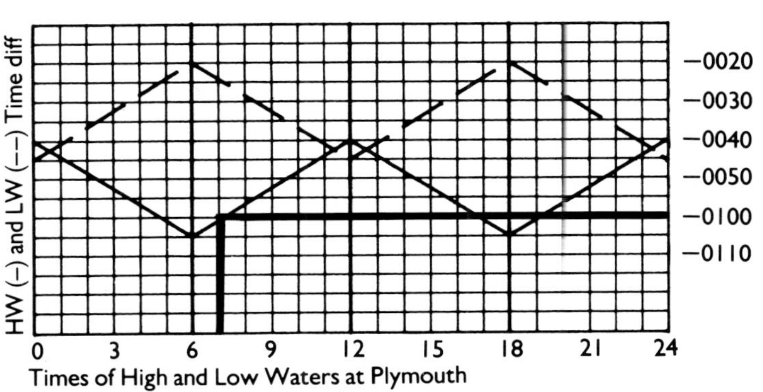

PILOT PACK

Volume 3

Bridport

See Chart No 301.

TIDE

Time and height differences on Standard Port (for instructions in use, see p. xii)

	Lat	Long	Time Difference HW		LW		Height Difference (metres)			
PLYMOUTH See p. 28	50°22′N	4°11′W	0100 and 1300	0600 and 1800	0100 and 1300	0600 and 1800	5.5	4.4	2.2	0.8
Bridport	50°42′N	2°45′W	+0025	+0040	0000	0000	−1.4	−1.4	−0.6	−0.2

TIDES relative to HW Dover (approx) HW −0500. Stream sets E +0510; W HW.

GENERAL

Bridport is actually about a mile inland and its port is West Bay which has a bar within the entrance but an area inside which has up to 4.6m. Good shelter in SW winds and reasonable when winds between SE and E. Some facilities.

APPROACH

There is a leading line, 10°, between the flagstaff, or Iso R 2s light, and the tower of Bridport church, about a mile inland. Enter −2HW+2. The entry signals are: by day, a R flag with a W St Andrew's cross; by night, the pier head lights are lit, but only for commercial vessels. A black ball indicates that the entrance is not clear. The entrance is only 12m wide and 180m long. Berth as directed by the Harbour Master, either against the quay, to starboard, or round the corner. It is possible to anchor off, but keep clear of the leading line and the sewer pipe.

CAUTIONS

1 The entrance is impossible in strong onshore winds and must not be attempted.
2 Do not attempt to enter at night.

TELEPHONES

Harbour Master:	Bridport 23222
HM Coastguard:	Portland 820441
MRSC:	Portland 820441
HM Customs:	Freephone Customs Yachts or (03057) 71189
Medical:	Bridport 23771 (Doctor)

VHF

None

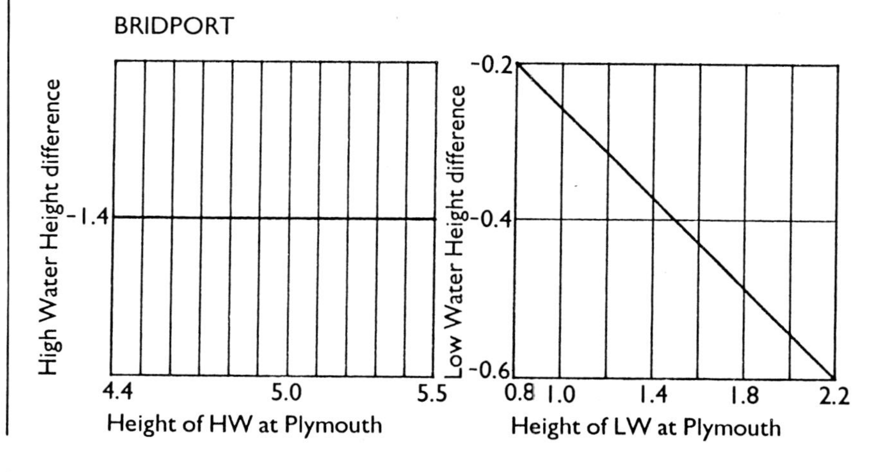

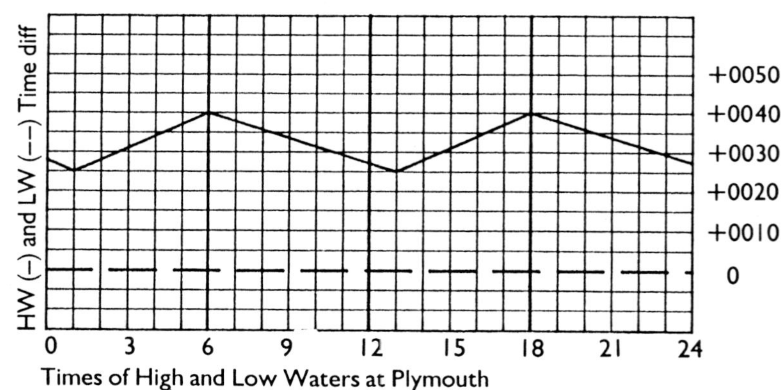

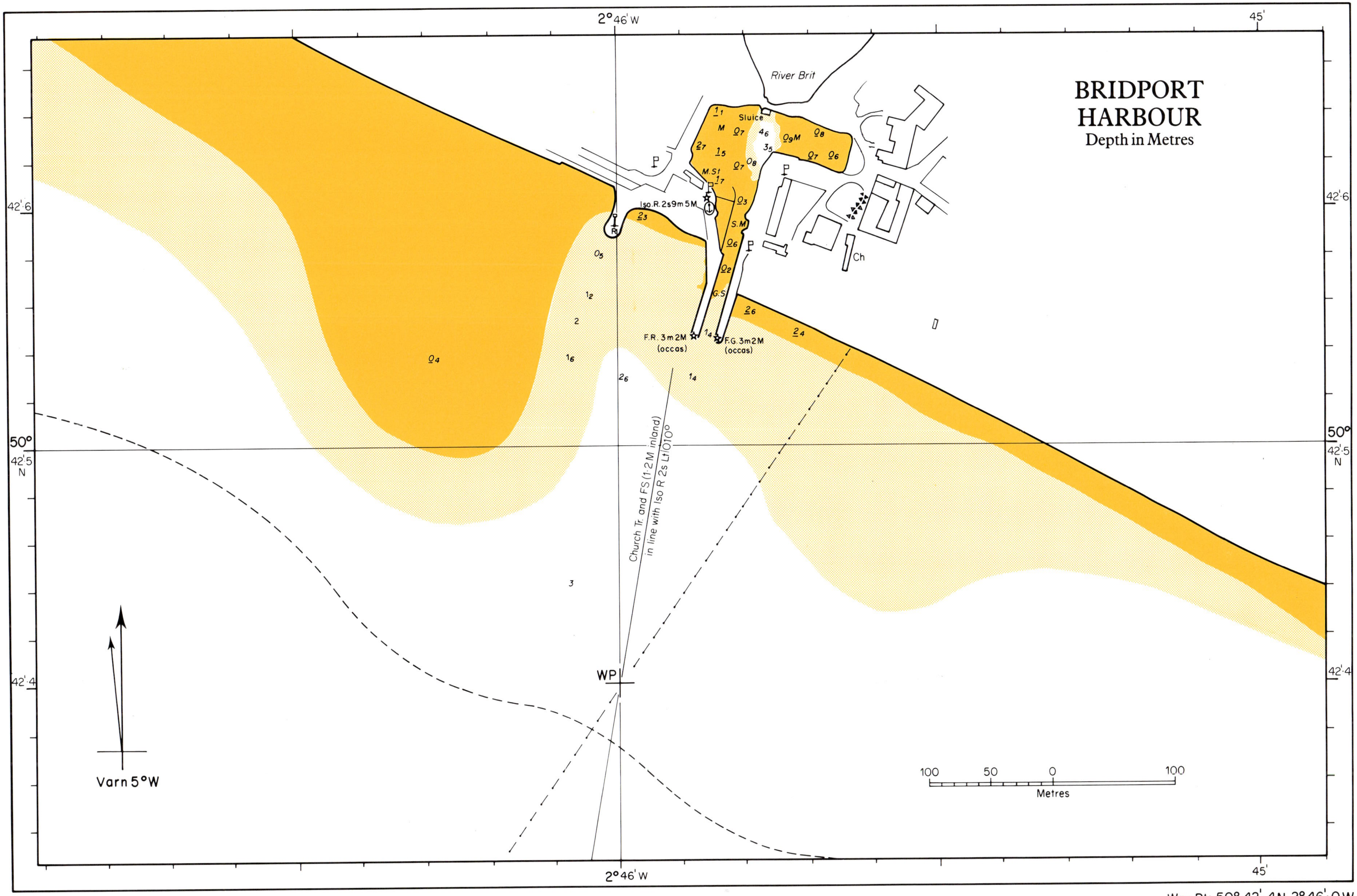

Chart No. 301

BRIDPORT HARBOUR

Depth in Metres

River Brit

Sluice

M.S1

S.M

G.S

Ch

Iso.R.2s9m5M

F.R.3m2M (occas)

F.G.3m2M (occas)

Church Tr. and FS (1·2M inland) in line with Iso R 2s Lt 010°

WP

Varn 5°W

100 50 0 100 Metres

2°46'W

45'

42'·6

50° 42'·5 N

42'·4

Way Pt: 50° 42'·4N 2°46'·0W

Lyme Regis

See Chart No 302.

TIDE

Time and height differences on Standard Port (for instructions in use, see p. xii)

	Lat	Long	Time Difference HW		LW		Height Difference (metres)			
			0100 and 1300	0600 and 1800	0100 and 1300	0600 and 1800				
PLYMOUTH See p. 28	50°22′N	4°11′W					5.5	4.4	2.2	0.8
Lyme Regis	50°43′N	2°56′W	+0040	+0100	+0005	−0005	−1.2	−1.3	−0.5	−0.2

TIDES relative to HW Dover (approx) HW −0445. Stream sets E +0450; W −0110.

GENERAL

Lyme Regis is a small holiday resort with a drying harbour. Shelter is good in prevailing SW winds and reasonable in SE and E winds. Some facilities.

APPROACH

There are leading lights. It is advisable to arrange a berth with the Harbour Master before entering as space is very limited. There are 12 berths for visiting craft on Victoria Pier. If it is decided to anchor off, take advice, if possible, as holding varies.

CAUTIONS

1 Do not attempt to enter in strong onshore winds.
2 There are many fishing floats in the area.

TELEPHONES

Harbour Master:	Lyme Regis 2137
MRSC:	Portland 820441
HM Customs:	Freephone Customs Yachts or (03057) 71189
Medical:	Lyme Regis 2696 or 2263 (Doctor)

VHF

Lyme Regis Radio:	Ch 16; 14

YACHT CLUBS

Lyme Regis SC:	Lyme Regis 3573
Lyme Regis Power Boat Club:	Lyme Regis 3788

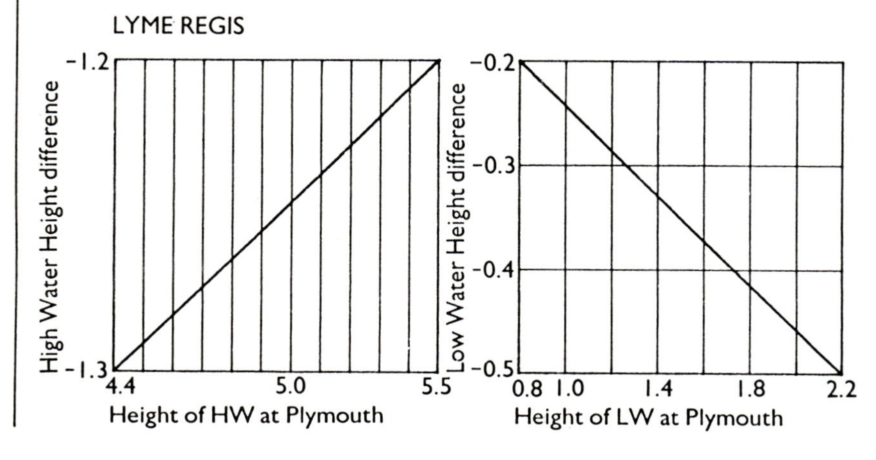

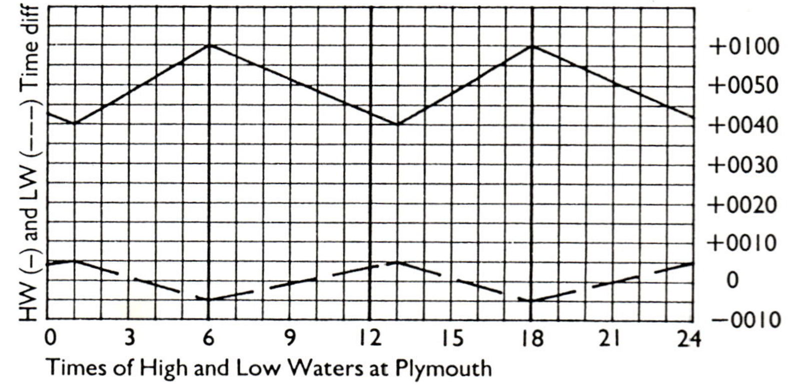

Chart No. 302

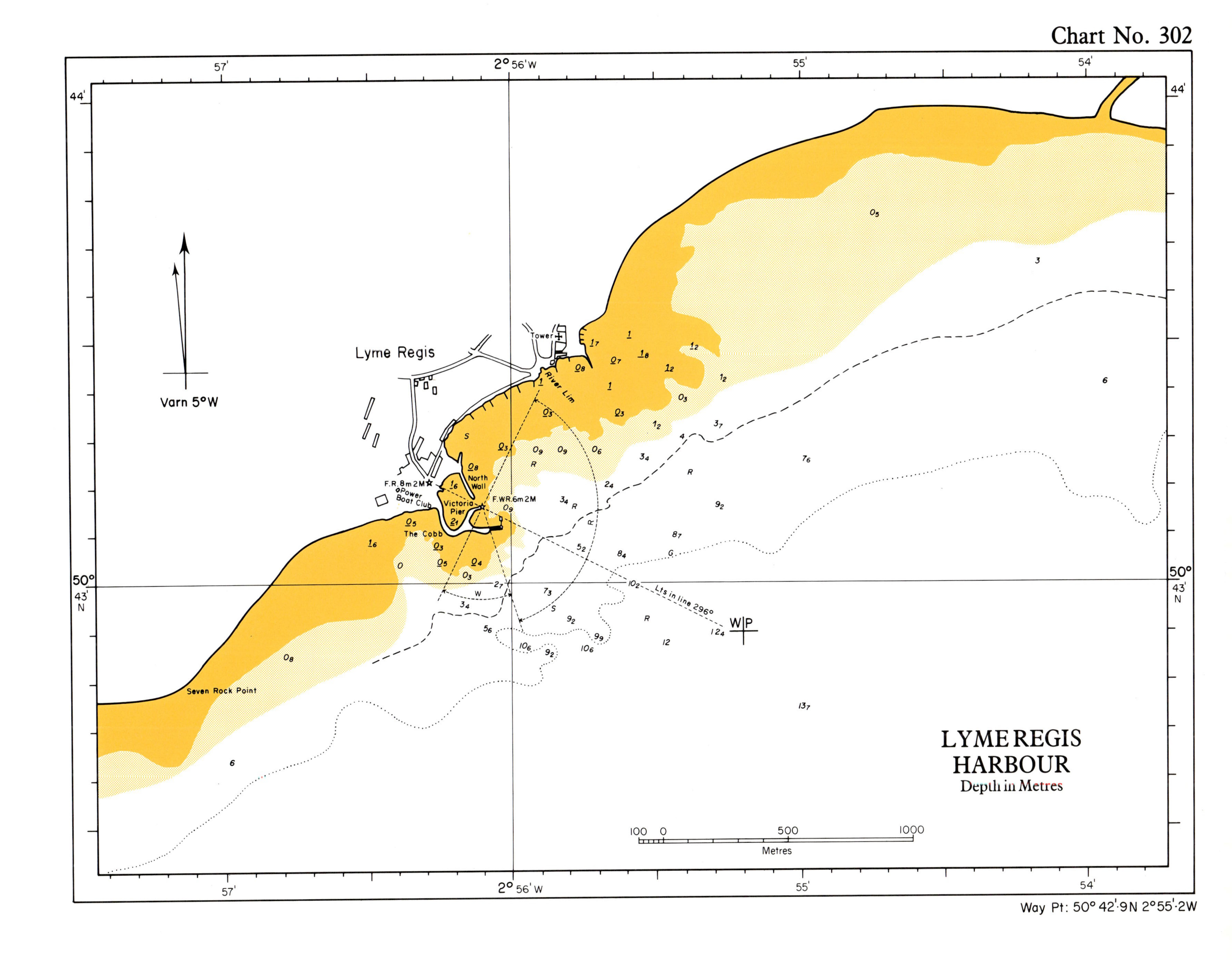
LYME REGIS
HARBOUR
Depth in Metres
Lyme Regis
Tower
River Lim
North Wall
Victoria Pier
The Cobb
F.R.8m2M
F.WR.6m2M
Power Boat Club
Seven Rock Point
Lts in line 296°
WP
Varn 5°W
100 0 500 1000
Metres
2° 56'W
57'
55'
54'
44'
50° 43' N
Way Pt: 50° 42'·9N 2°55'·2W

Exmouth

See Chart No 303.

TIDE

Time and height differences on Standard Port (for instructions in use, see p. xii)

	Lat	Long	Time Difference HW		Time Difference LW		Height Difference (metres)			
			0100 and 1300	0600 and 1800	0100 and 1300	0600 and 1800				
PLYMOUTH See p. 28	50°22′N	4°11′W					5.5	4.4	2.2	0.8
Exmouth Apprs	50°36′N	3°23′W	+0030	+0050	+0015	+0005	−0.9	−1.0	−0.5	−0.3
Exmouth Dock	50°37′N	3°25′W	+0040	+0100	+0050	+0020	−1.5	−1.6	−0.9	−0.6

TIDES relative to HW Dover (approx) HW −0455. Stream sets E +0445; W 0000.

GENERAL

Exmouth is a holiday resort with a small tidal dock approached via a swing bridge with the possibility of finding a mooring or space to anchor in the river. There is a canal up to Exeter – see page 8. All facilities.

APPROACH

The approach is clear of dangers and then, from Exe Fairway buoy, is marked by buoys numbered from seaward, odd to starboard and even to port, to suit the shifting channel. The entrance is perfectly safe, given water and fine weather, but should not be used in strong onshore winds.

CAUTION

1 See Approach above.

TELEPHONES

Dock Master: Exmouth 272009
Harbour Master: Exeter 74306
HM Coastguard: Exmouth 263232
MRSC: Brixham 58292
HM Customs: Freephone Customs Yachts or (0752) 669811
Medical: Exmouth 273001 (Doctor)

VHF

Exeter Harbour — Ch 16; 6; 12

YACHT CLUB

Exe SC: Exmouth 264607

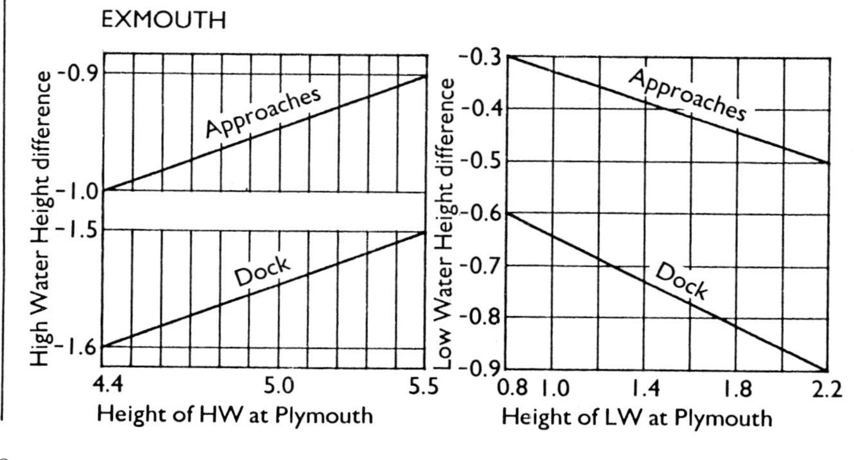

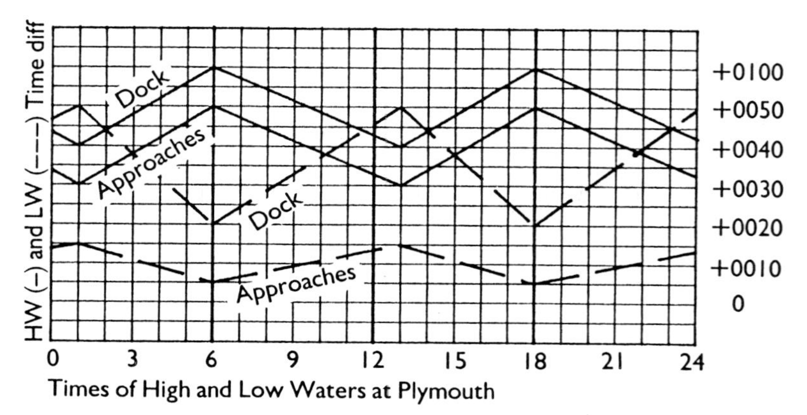

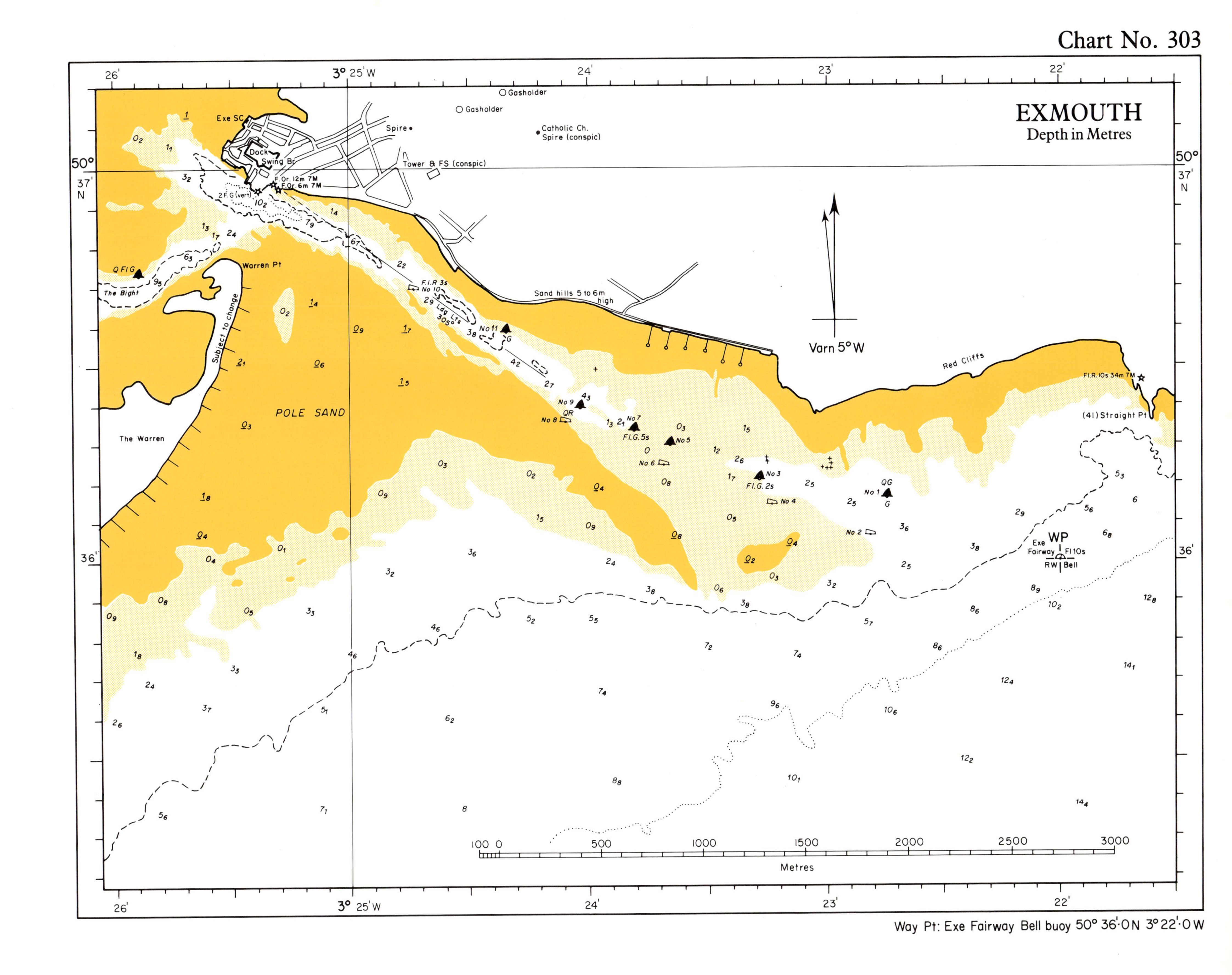

Way Pt: Exe Fairway Bell buoy 50° 36'·0N 3° 22'·0W

River Exe

See Chart No 304.

TIDE

Time and height differences on Standard Port (for instructions in use, see p. xii)

	Lat	Long	Time Difference HW		LW		Height Difference (metres)			
			0100 and 1300	0600 and 1800	0100 and 1300	0600 and 1800				
PLYMOUTH See p. 28	50°22′N	4°11′W					5.5	4.4	2.2	0.8
Starcross	50°38′N	3°27′W	+0040	+0110	no data		−1.4	−1.5	no data	
Topsham	50°54′N	3°28′W	+0045	+0105	no data		−1.5	−1.6	no data	

GENERAL

The river winds and has many drying banks and patches. It gives access to Exeter via a canal entered at Turf Lock which has one further lock and four opening bridges. Anchor, if convenient swinging space can be found clear of the fairway and any moorings, off the Point in 3m, off Bull Hill Bank. It may be possible to pick up a mooring, with permission, off Starcross.

APPROACH

See page 6. For a berth or moorings, apply to the Harbour Master.

TELEPHONE

Harbour Master — Exeter 74306

VHF

Exeter Harbour: — Ch 16; 6; 12

YACHT CLUBS

Starcross YC: — Starcross 890470

Starcross Fishing & Cruising Club: — Starcross 890582

Topsham SC:

RIVER EXE

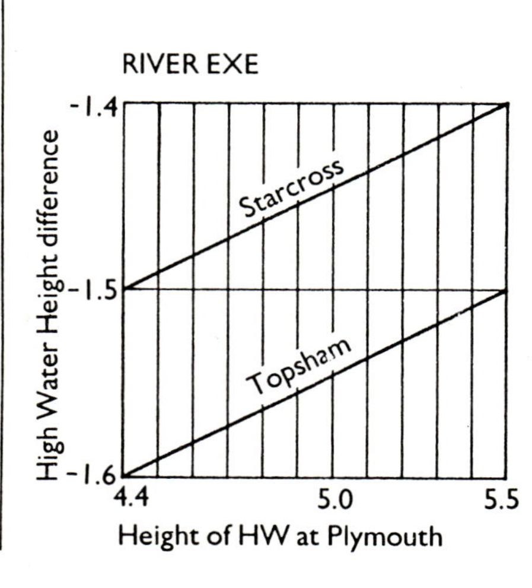

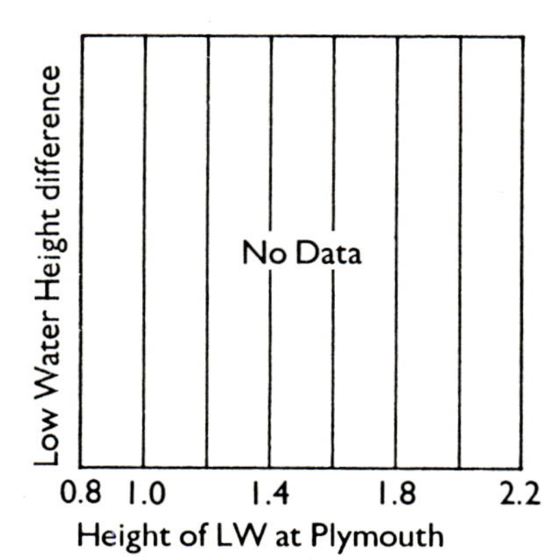

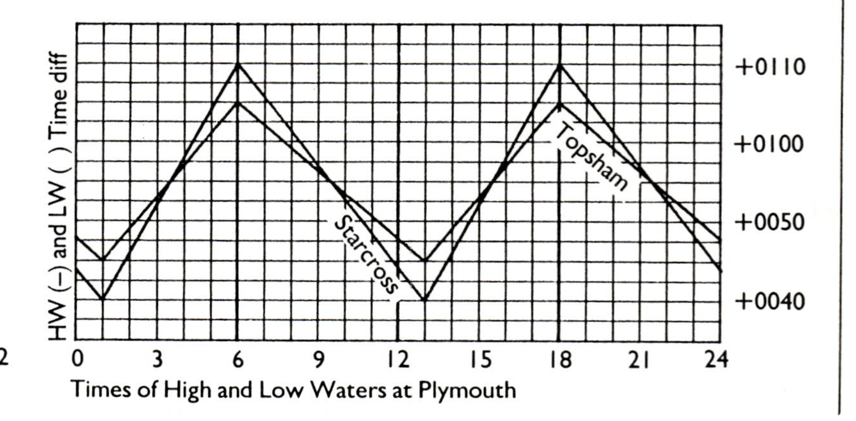

Chart No. 304

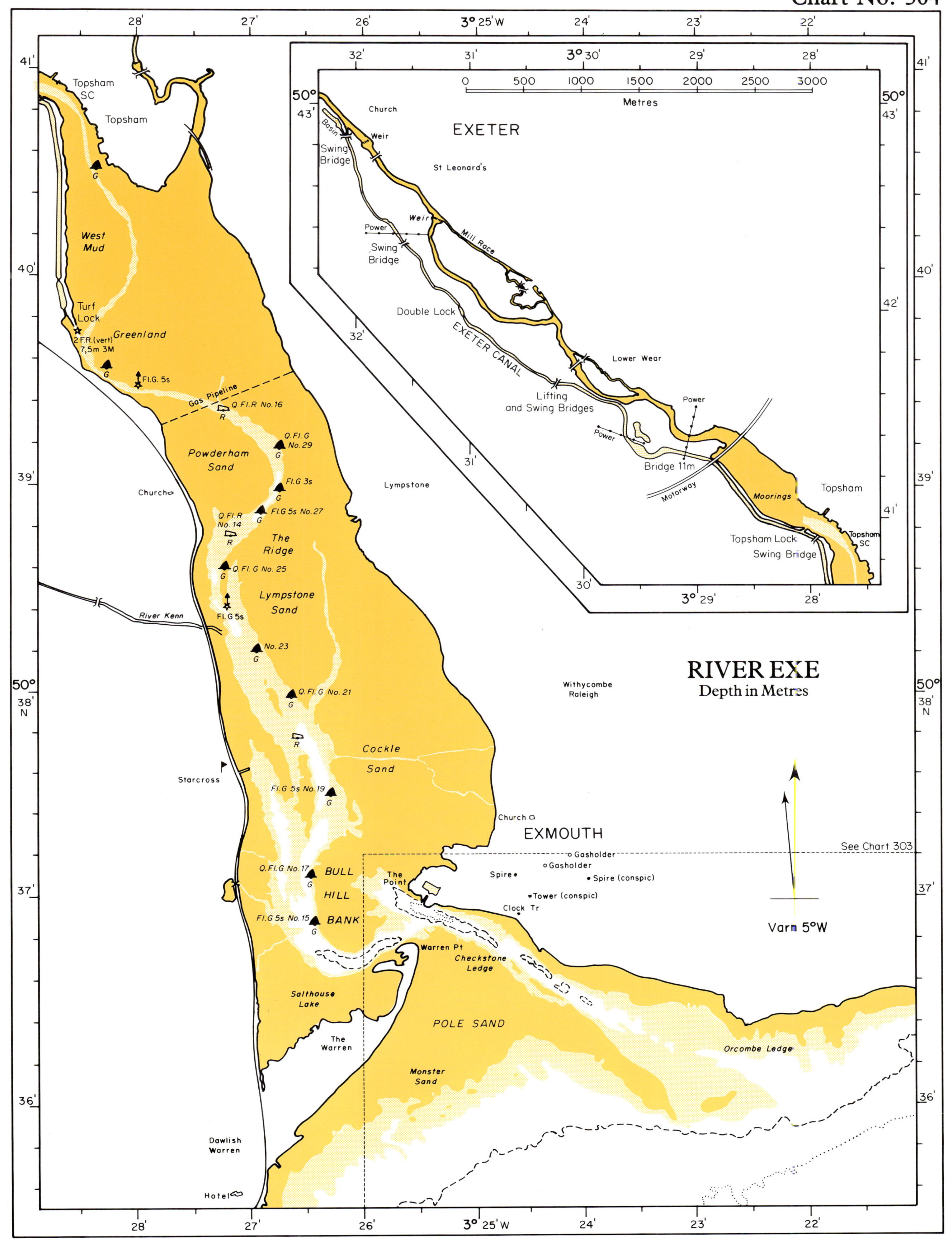

RIVER EXE
Depth in Metres
EXETER
Church
Basin
Weir
Swing Bridge
St Leonard's
Weir
Power
Mill Race
Swing Bridge
Double Lock
EXETER CANAL
Lower Wear
Lifting and Swing Bridges
Power
Power
Bridge 11m
Motorway
Moorings
Topsham
Topsham Lock
Swing Bridge
Topsham SC
Metres
Topsham SC
Topsham
West Mud
Turf Lock
2 F.R.(vert) 7,5m 3M
Greenland
Fl.G. 5s
Gas Pipeline
Q.Fl.R No.16
Q.Fl.G No. 29
Powderham Sand
Fl.G 3s
Church
Lympstone
Q.Fl.R No. 14
Fl.G 5s No. 27
The Ridge
Q.Fl.G No. 25
Lympstone Sand
River Kenn
Fl.G 5s
No. 23
Q.Fl.G No. 21
Withycombe Raleigh
Cockle Sand
Starcross
Fl.G 5s No. 19
Church
EXMOUTH
See Chart 303
Gasholder
Gasholder
Spire
Spire (conspic)
Tower (conspic)
Clock Tr
Q.Fl.G No. 17
BULL HILL BANK
The Point
Fl.G 5s No. 15
Warren Pt
Checkstone Ledge
Varn 5°W
Salthouse Lake
POLE SAND
The Warren
Orcombe Ledge
Monster Sand
Dawlish Warren
Hotel

Teignmouth

See Chart No. 305.

TIDE

Time and height differences on Standard Port (for instructions in use, see p. xii)

	Lat	Long	Time Difference HW		LW		Height Difference (metres)			
			0100 and 1300	0600 and 1800	0100 and 1300	0600 and 1800				
PLYMOUTH See p. 28	50°22′N	4°11′W					5.5	4.4	2.2	0.8
Teignmouth Apprs	50°33′N	3°30′W	+0025	+0040	0000	0000	−0.7	−0.8	−0.3	−0.2

TIDES relative to HW Dover (approx) HW −0500. Stream sets (outside the bar) SSW −0030; NNE +0510; (in the harbour entrance) Floods +0110; Ebbs −0500.

GENERAL

Teignmouth is a holiday resort with some commercial shipping. Unfortunately, there is little room for visiting craft and the approach (see below) is also discouraging.

APPROACH

Teignmouth faces E across Lyme Bay and has a clear offing until the bar and sands at the entrance are encountered. These shift constantly and, though marked with buoys, require local knowledge unless there is plenty of water and fine weather. The best water may be over the Bar or over either bank of sand. Once inside, the training wall scours deep water on the S side and then the channel swings sharp N to the quays and then W to the bridge, which has a drawbridge for small vessels wishing to proceed to Newton Abbot. All facilities. Do not anchor within the harbour but apply to the Harbour Master for a mooring.

CAUTION

1 See notes on Approach above.

TELEPHONES

Harbour Master:	Teignmouth 2311
HM Coastguard:	Brixham 58292
MRSC:	Brixham 58292
HM Customs	Freephone Customs Yachts or (0752) 669811
Medical:	Teignmouth 4355 (Doctor) or 2161 (Hospital)

VHF

Teignmouth Harbour: Ch 16; 12 (office hours)

YACHT CLUB

Teign Corinthian YC: Teignmouth 2734

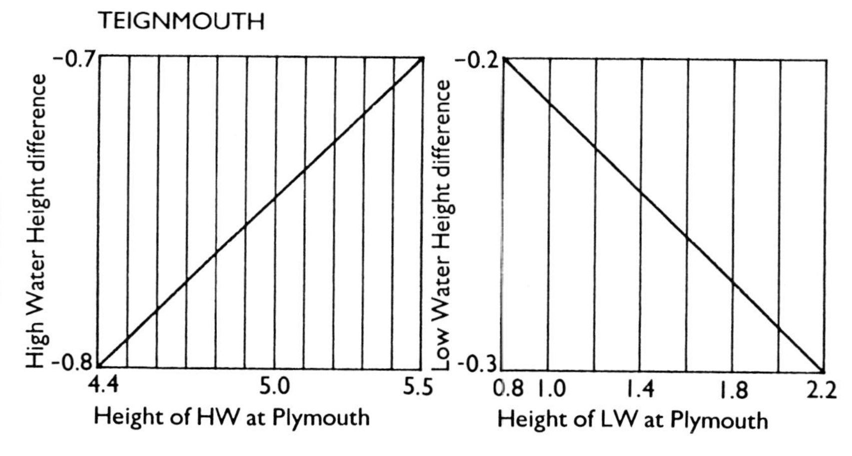

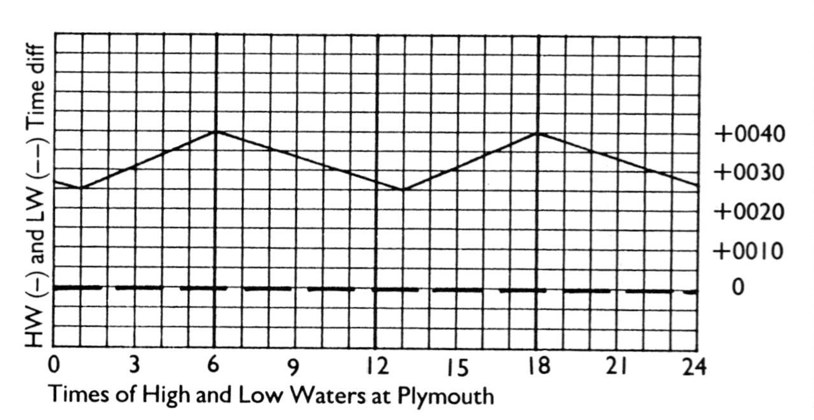

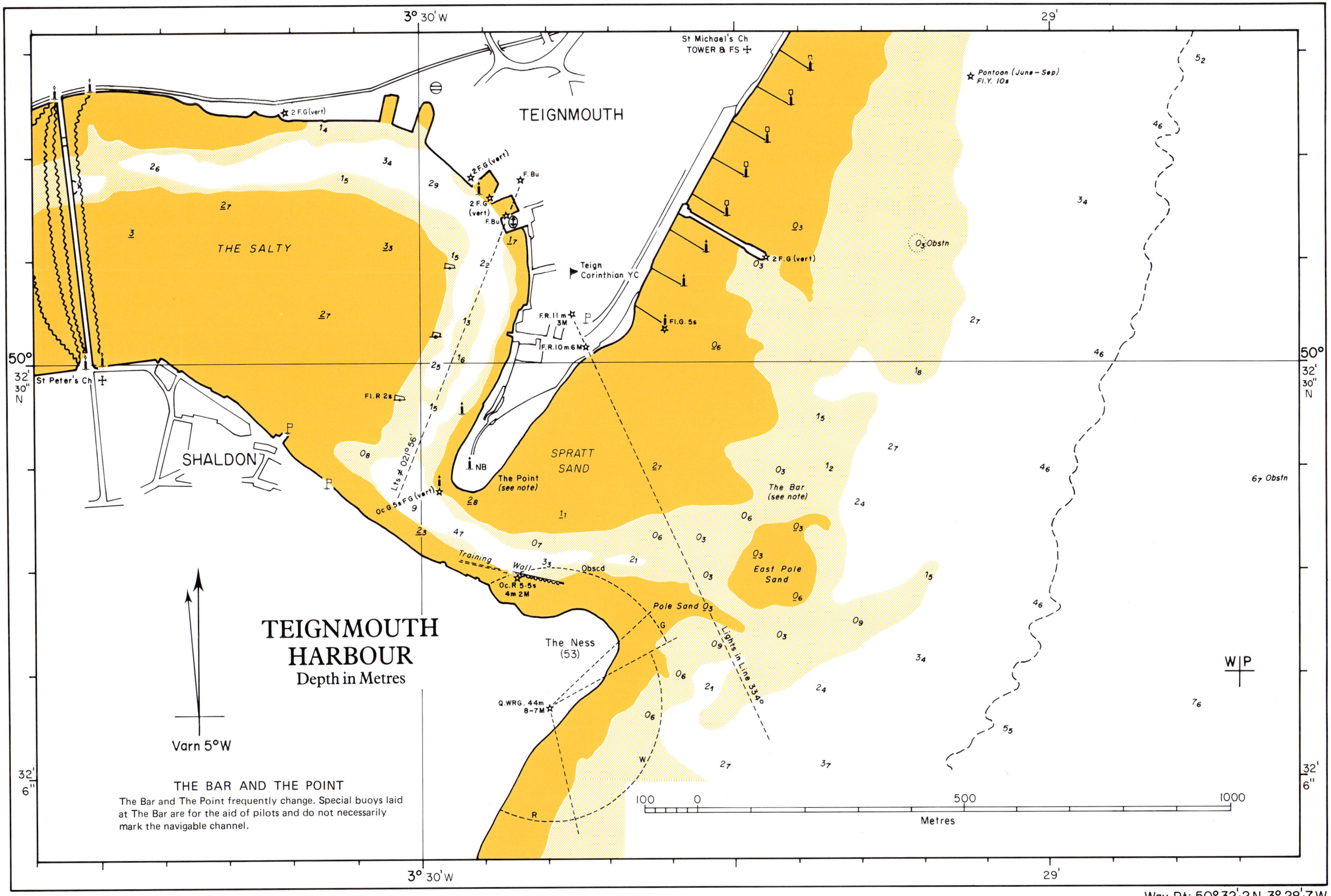
TEIGNMOUTH
HARBOUR
Depth in Metres
THE BAR AND THE POINT
The Bar and The Point frequently change. Special buoys laid at The Bar are for the aid of pilots and do not necessarily mark the navigable channel.
Varn 5°W
TEIGNMOUTH
SHALDON
THE SALTY
SPRATT SAND
East Pole Sand
Pole Sand
The Ness (53)
The Point (see note)
The Bar (see note)
St Michael's Ch TOWER & FS
St Peter's Ch
Teign Corinthian YC
Pontoon (June – Sep) Fl.Y. 10s
2 F.G (vert)
F.Bu
F.R. 11 m 3M
F.R. 10 m 6M
Fl.G. 5s
Fl.R 2s
Oc G 5s F.G (vert)
Lts ≠ 021°56'
Lights in Line 334°
Training Wall
Oc.R 5·5s 4m 2M
Q.WRG. 44m 8-7M
Obscd
O3 Obstn
67 Obstn
NB
W|P
100 0 500 1000
Metres
3° 30' W
29'
50° 32' 30" N
32' 6"

Way Pt: 50° 32'·2 N 3° 28'·7 W

Torquay

See Chart No 306.

TIDE

Time and height differences on Standard Port (for instructions in use, see p. xii)

	Lat	Long	Time Difference HW		LW		Height Difference (metres)			
			0100 and 1300	0600 and 1800	0100 and 1300	0600 and 1800				
PLYMOUTH See p. 28	50°22′N	4°11′W					5.5	4.4	2.2	0.8
Torquay	50°28′N	3°31′W	+0025	+0045	+0010	0000	−0.6	−0.7	−0.2	−0.1

TIDES relative to HW Dover (approx) HW −0500. Stream sets W −0100; E +0500.

GENERAL

Torquay is a holiday and yachting centre with a large new marina and good shelter in all except strong S winds. All facilities.

APPROACH

The approach holds no dangers. Three R balls or lts indicate the harbour is closed. There is good holding off the harbour. A controlled area up to about 2 cables off the beach and marked with Y spherical buoys is for swimming and there is a speed limit of 5 knots within it.

CAUTIONS

1 See the note about the controlled area in Approach above.
2 The entrance may be busy with tripper boats.

TELEPHONES

Harbour Master:	Torquay 22429
HM Coastguard:	Brixham 58292
MRSC:	Brixham 58292
HM Customs:	Freephone Customs Yachts or (0752) 669811
Medical:	Torquay 64567 (Hospital)

VHF

Torquay Harbour:	Ch 14; 16 (office hours)
Torquay Marina:	Ch M

MARINA

Torquay Marina:	Torquay 214624

YACHT CLUB

R Torbay YC:	Torquay 22006

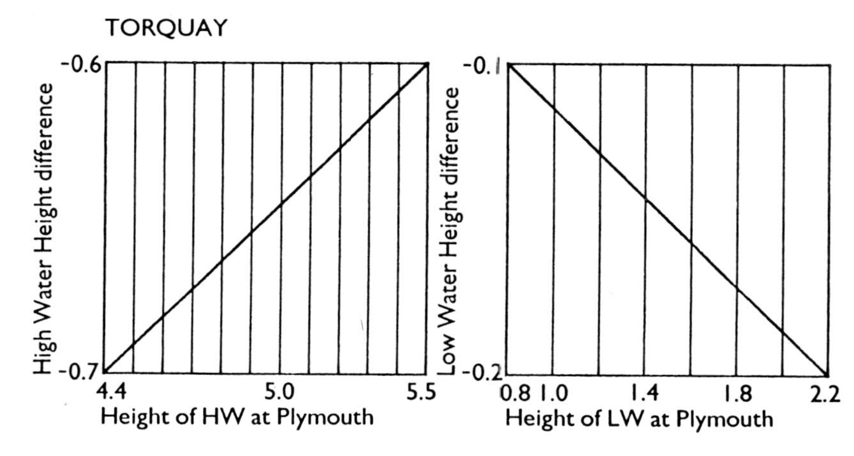

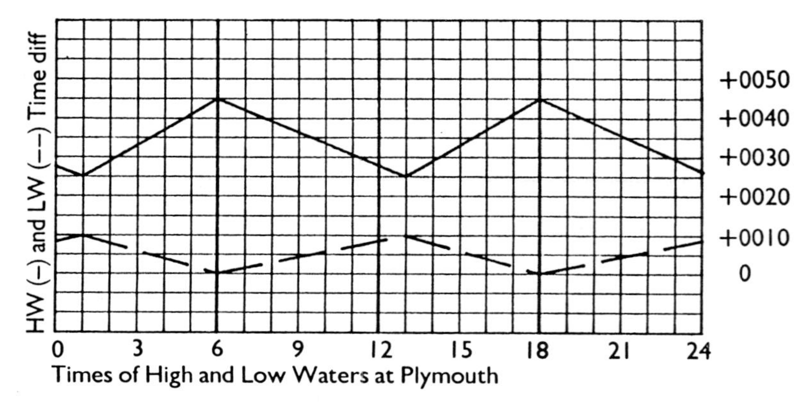

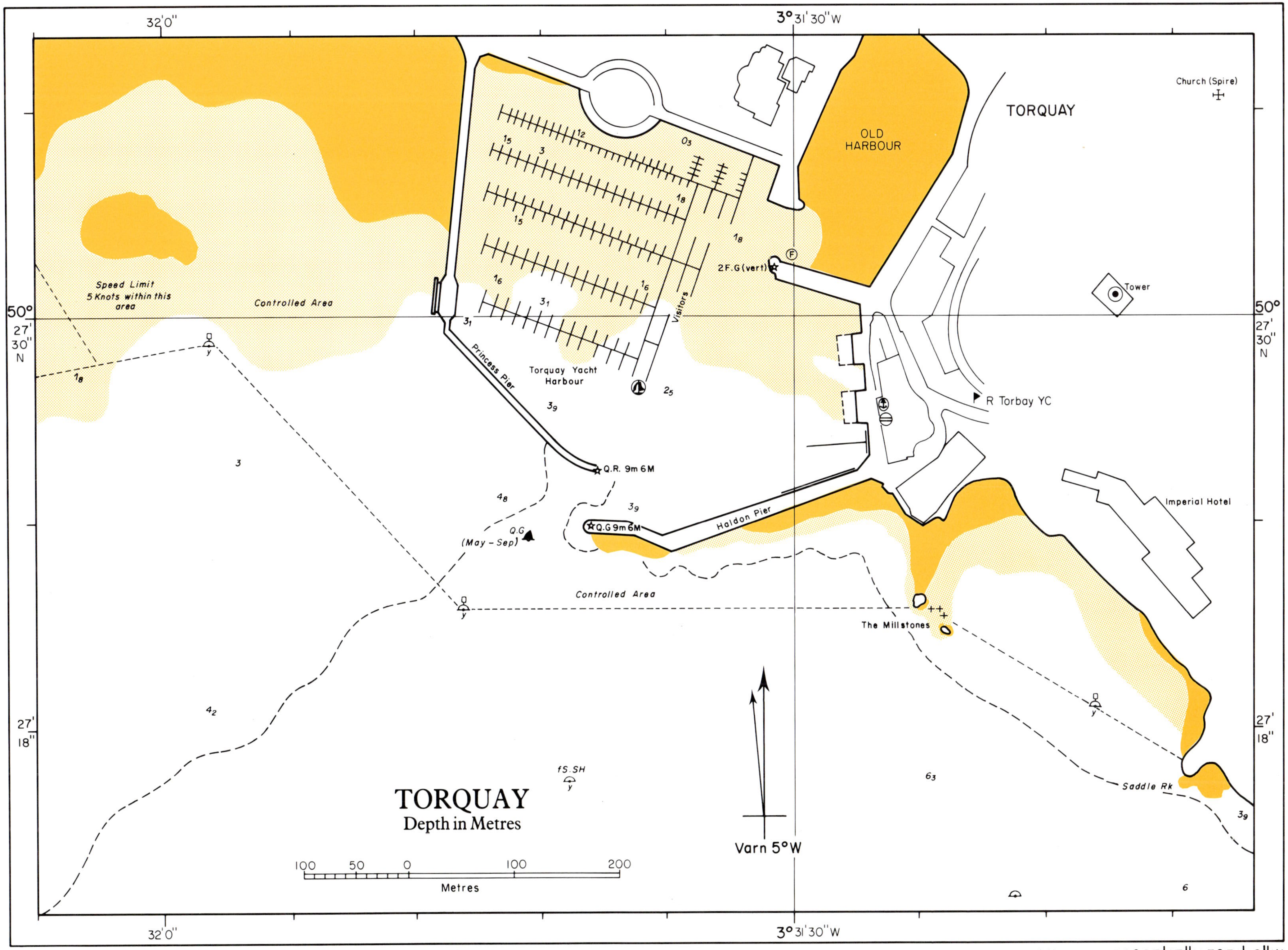

Chart No. 306

TORQUAY
Depth in Metres

Way Pt: Saddle Rock 50°27'17"N 3°31'12"W

32'0"
3°31'30"W
50° 27' 30" N
27' 18"

TORQUAY
OLD HARBOUR
Church (Spire)
Tower
R Torbay YC
Imperial Hotel
2F.G(vert)
Visitors
Torquay Yacht Harbour
Princess Pier
Q.R. 9m 6M
Q.G 9m 6M
Haldon Pier
Q.G (May – Sep)
Controlled Area
Speed Limit 5 Knots within this area
The Millstones
Saddle Rk
fS.SH
Varn 5°W

100 50 0 100 200
Metres

Brixham

See Chart No 307.

TIDE

See Torquay on page 12.

TIDES relative to HW Dover (approx) HW −0505. Stream sets SW −0100; NE +0500.

GENERAL

Brixham is an important fishing port and yachting centre. There is a new (1989), large marina with space reserved for visiting yachts on the wave screen. All facilities.

APPROACH

The approach is clear of all dangers but, as the harbour is congested and there are no leading lights, entrance should not be attempted at night without local knowledge.

CAUTION

1 Strong NW winds make the outer harbour dangerous.

TELEPHONES

Harbour Master	Brixham (08045) 3321
HM Coastguard:	Brixham (0803) 882704
MRSC:	Brixham (0803) 882704
HM Customs:	Freephone Customs Yachts or (0752) 669811
Medical:	Brixham (08045) 2731 (Doctor)

VHF

Brixham Harbour	Ch 14; 16

MARINA

Prince William Marina:	Brixham (0803) 882711

YACHT CLUB

Brixham YC:	Brixham (08045) 3332

Chart No. 307

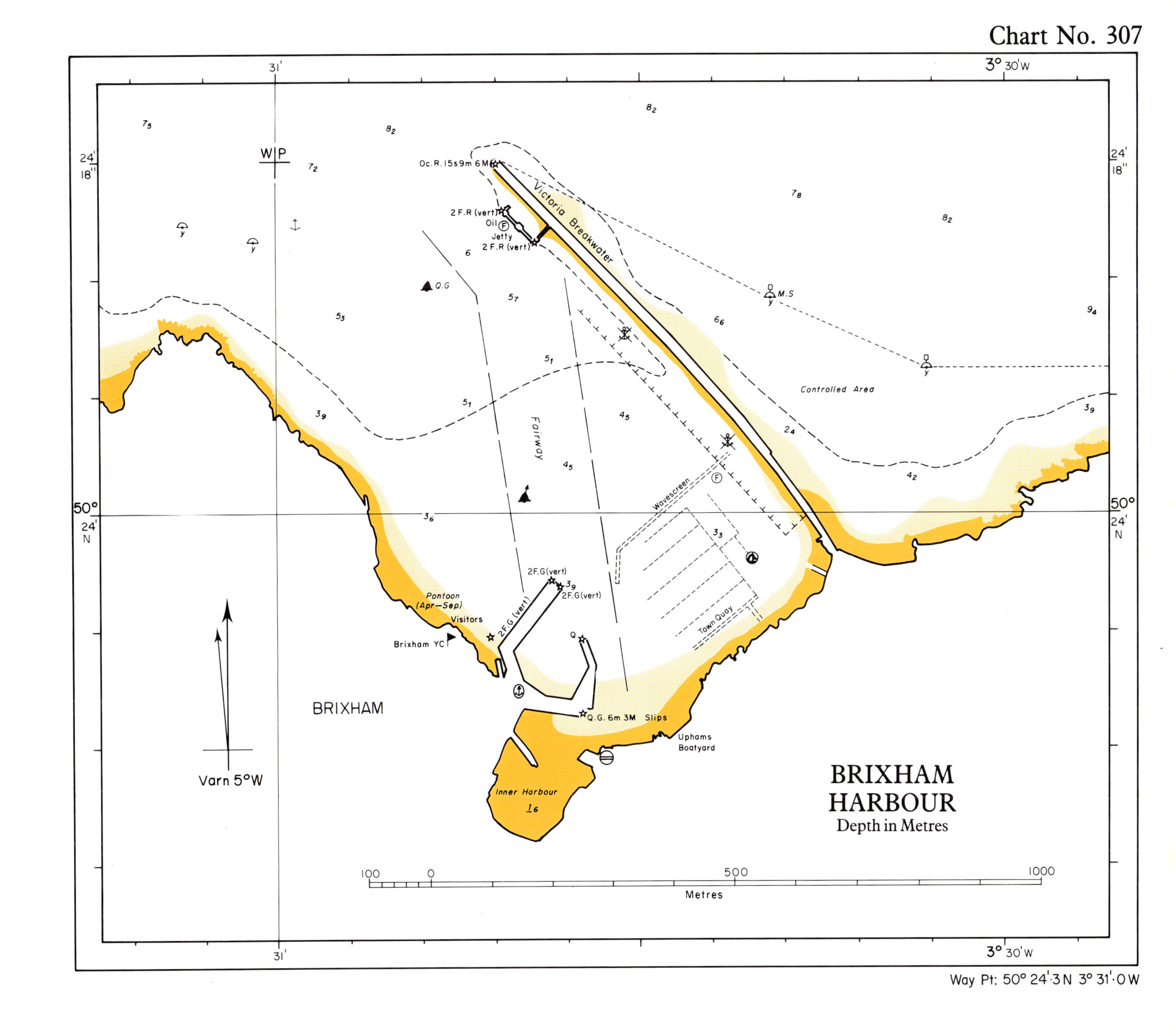
BRIXHAM HARBOUR
Depth in Metres
Victoria Breakwater
Oc.R.15s9m 6M
2 F.R (vert)
Oil
Jetty
2 F.R (vert)
Q.G
M.S
Controlled Area
Fairway
Wavescreen
Town Quay
2F.G(vert)
2F.G(vert)
2.F.G (vert)
Q
Q.G. 6m 3M
Slips
Uphams Boatyard
Pontoon (Apr–Sep)
Visitors
Brixham YC
BRIXHAM
Inner Harbour
W/P
Varn 5°W
Metres
Way Pt: 50° 24'.3N 3° 31'.0W

Dartmouth

See Chart No 308.

TIDE

Time and height differences on Standard Port (for instructions in use, see p. xii)

	Lat	Long	Time Difference HW		Time Difference LW		Height Difference (metres)			
PLYMOUTH See p. 28	50°22′N	4°11′W	0100 and 1300	0600 and 1800	0100 and 1300	0600 and 1800	5.5	4.4	2.2	0.8
Dartmouth	50°21′N	3°34′W	+0025	+0040	+0015	0000	−0.8	−0.6	−0.1	−0.2

TIDES relative to HW Dover (approx) HW −0515. Stream sets SW −0100; NE +0540.

GENERAL

The River Dart provides a deep water sheltered harbour with a safe approach in any weather. All facilities and three marinas. The river extends to Totnes, see page 18.

APPROACH

There is a prominent day mark to the E of the entrance and a sectored light giving a lead. It is possible to anchor off, but there is often a nasty swell. The marinas from seaward are: Darthaven (starboard), Dart (port) and Kingswear (starboard). The HM may be able to offer a mooring or it may be possible to come alongside the pontoons temporarily at the R Dart YC (starboard) or the Dartmouth YC (port). Anchoring within the river, even off the fairway, is not recommended owing to the presence of mooring chains on the bottom.

CAUTIONS

1 The area between the Mewstone and the land is littered with fishing lines.
2 Both sides of the entrance have dangerous rocks. Do not cut the corners.
3 The lower ferry has right of way.

TELEPHONES

Harbour Master: Dartmouth 2337
HM Coastguard: Brixham 58292
MRCC: Brixham 58292
HM Customs: Freephone Customs Yachts or (0752) 669811
Medical: Dartmouth 2212 (Doctor) or 2255 (Hospital)

VHF

Dartmouth Harbour: Ch 11
Darthaven Marina: Ch M
Dart Marina: Ch M
Kingswear Marina (call DART MARINA FOUR): Ch M

MARINAS

Darthaven: Kingswear 545
Dart: Dartmouth 3351 Ext 58
Kingswear: Dartmouth 3351 Ext 38

YACHT CLUBS

R Dart YC: Kingswear 272
The Dartmouth YC: Dartmouth 2305

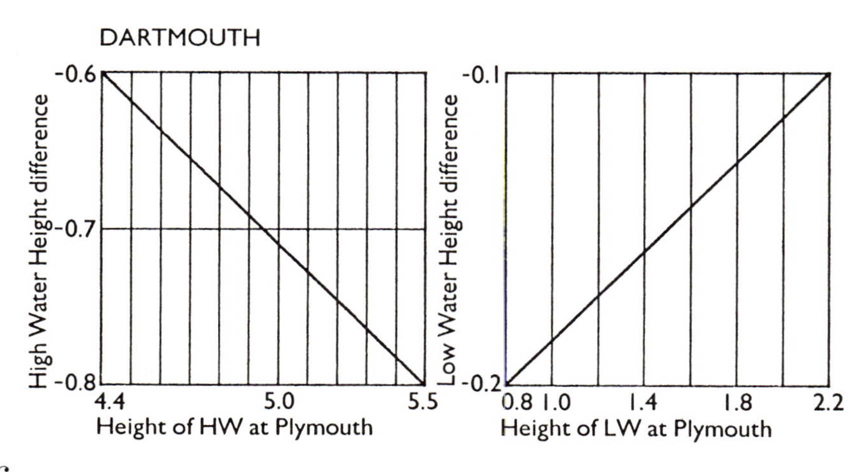

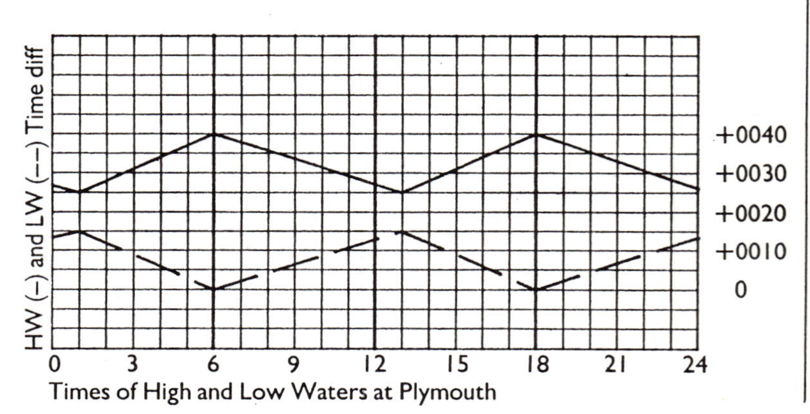

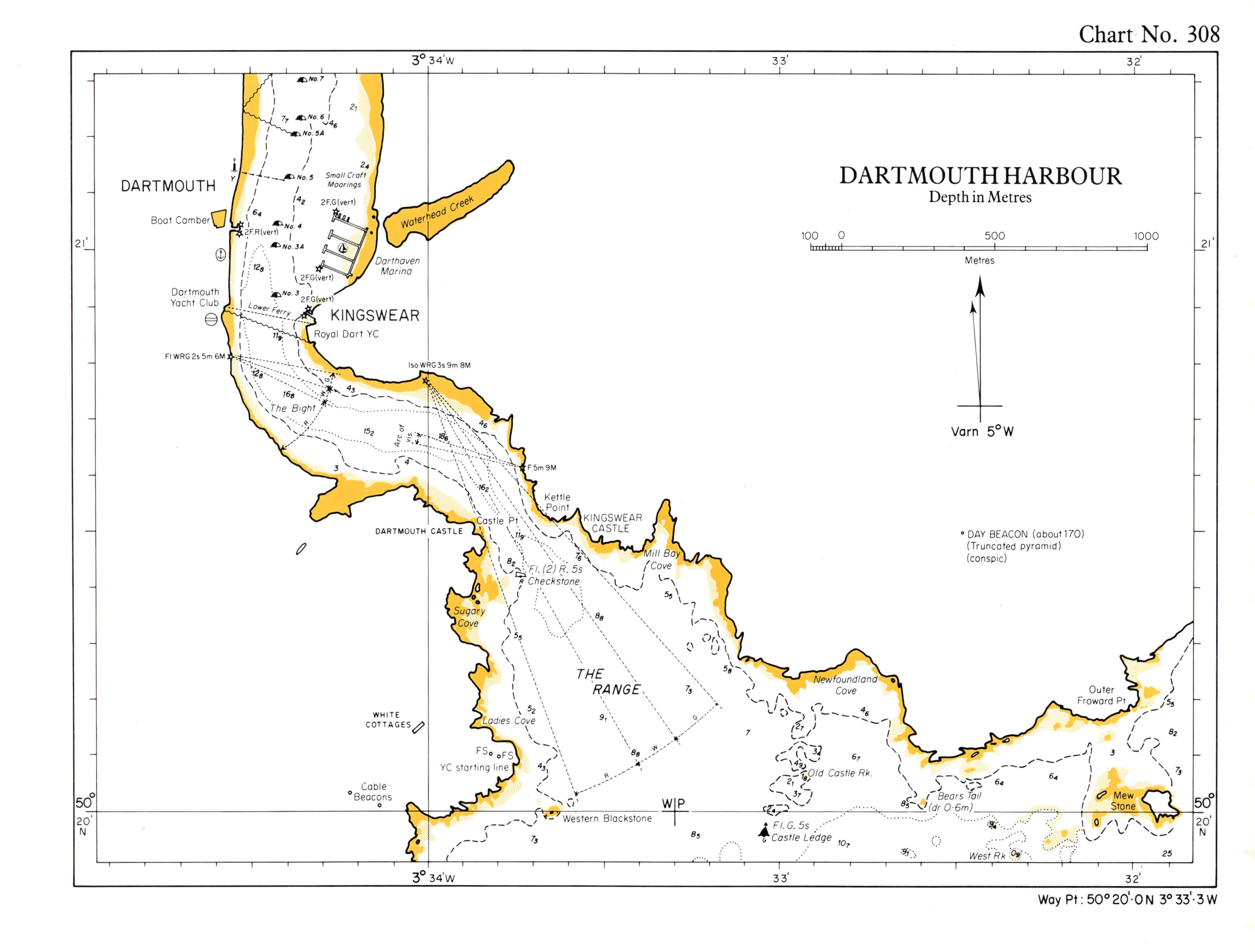
DARTMOUTH HARBOUR
Depth in Metres
100 0 500 1000
Metres
Varn 5°W
DARTMOUTH
Boat Camber
2F.R(vert)
Dartmouth Yacht Club
Fl WRG 2s 5m 6M
KINGSWEAR
Royal Dart YC
Small Craft Moorings
2F.G(vert)
Waterhead Creek
Darthaven Marina
Lower Ferry
The Bight
Iso WRG 3s 9m 8M
Arc of vis.
F 5m 9M
Kettle Point
KINGSWEAR CASTLE
DARTMOUTH CASTLE
Castle Pt
Fl.(2) R. 5s
Checkstone
Mill Bay Cove
Sugary Cove
THE RANGE
Ladies Cove
WHITE COTTAGES
FS FS
YC starting line
Cable Beacons
Western Blackstone
W/P
Fl. G. 5s
Castle Ledge
Old Castle Rk.
Newfoundland Cove
Bears Tail (dr 0·6m)
West Rk
Outer Froward Pt
Mew Stone
DAY BEACON (about 170)
(Truncated pyramid)
(conspic)
3° 34'W
33'
32'
21'
50° 20' N
Way Pt: 50° 20'·0N 3° 33'·3W

River Dart

See Chart No 309.

TIDE

Time and height differences on Standard Port (for instructions in use, see p. xii)

	Lat	Long	Time Difference HW		Time Difference LW		Height Difference (metres)			
			0100 and 1300	0600 and 1800	0100 and 1300	0600 and 1800	5.5	4.4	2.2	0.8
PLYMOUTH See p. 28	50°22′N	4°11′W								
Greenway Quay	50°23′N	3°35′W	+0030	+0045	+0025	+0005	−0.6	−0.6	−0.2	−0.2
Totnes	50°26′N	3°41′W	+0025	+0040	+0115	+0030	−2.0	−2.1	dries	

GENERAL

The River Dart is very beautiful and navigable as far as Totnes on the tide. Quite large coasters use the river but craft should not stay at Totnes beyond HW unless prepared to dry out. The river is well marked with buoys and beacons and, generally, the deepest water is on the outside of the bends. Craft may anchor off Parsons Mud just S of Vipers Quay or should consult the River Officer who is usually afloat near Dittisham.

APPROACH

See page 16.

CAUTIONS

1 The Anchor Stone off Vipers Quay and the Eel Rock 110m SE of Waddeton boat house, both of which dry, must be noted and avoided.
2 Though there is No 1 (R can) buoy in the middle of the river at Galmpton, yachts should follow the line of moorings where there is deeper water.
3 The ebb is much stronger than the flood, especially after heavy rain when, near Totnes, there may not be a flood stream at all.

TELEPHONES

See page 16.

VHF

See page 16.

MARINAS

See page 16.

YACHT CLUBS

See page 16.

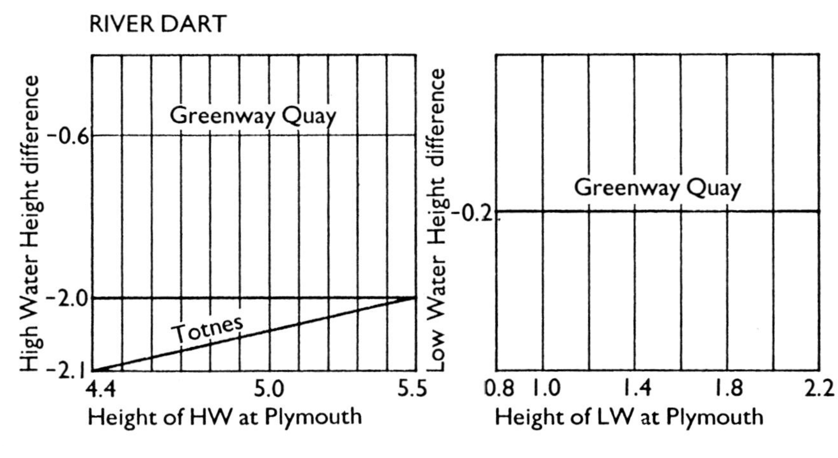

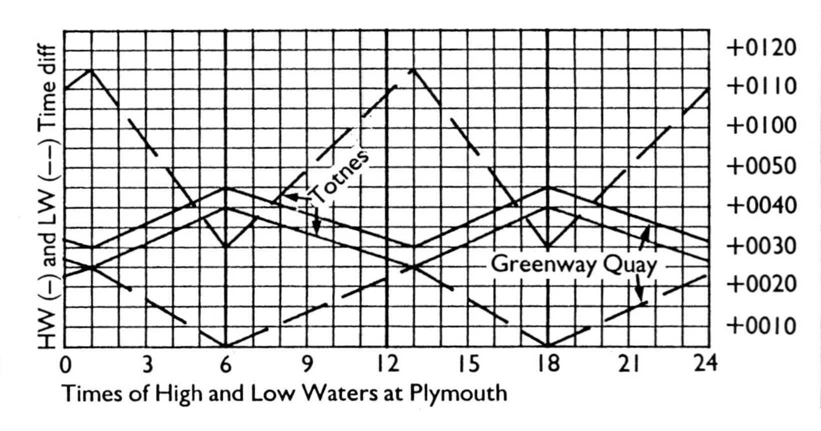

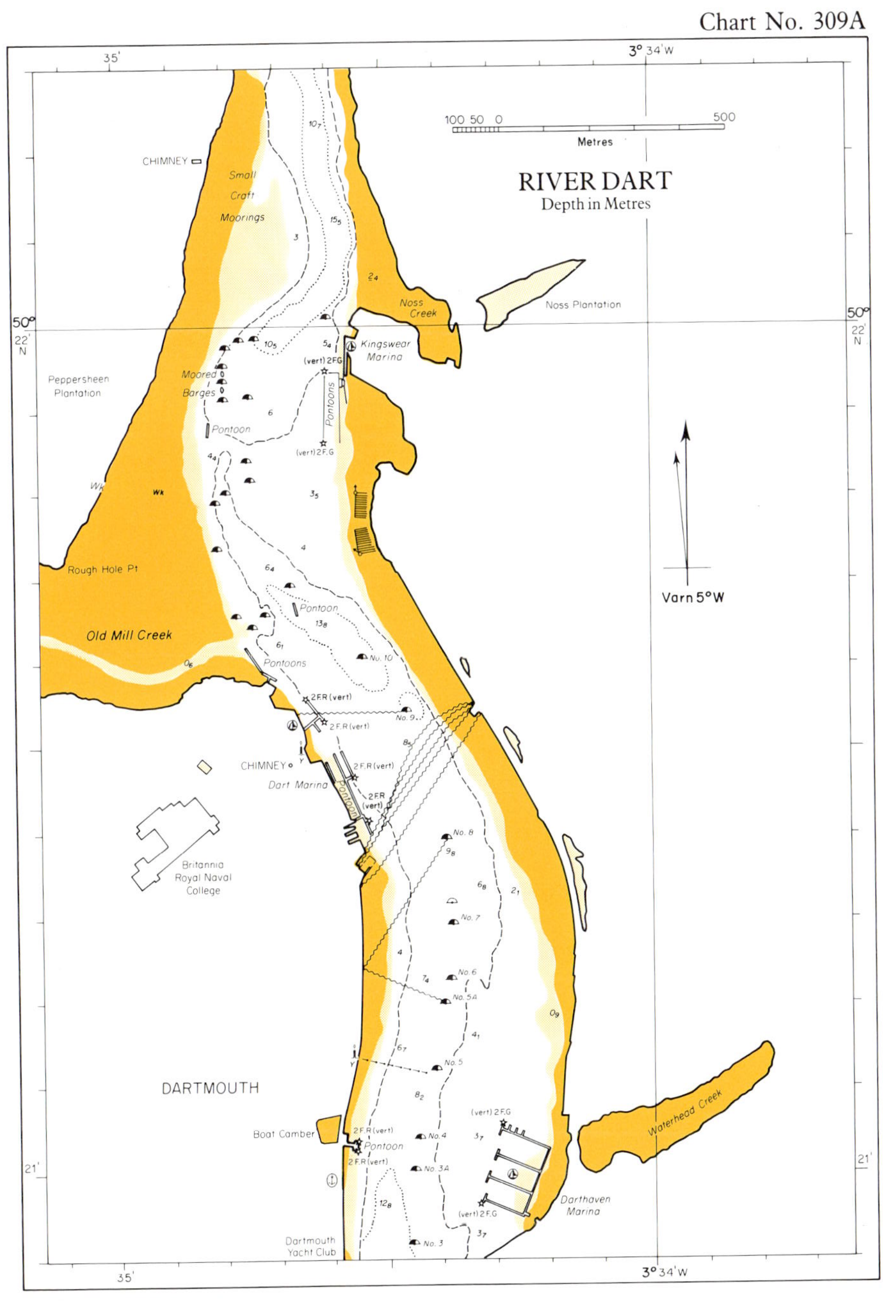
Chart No. 309A
RIVER DART
Depth in Metres
Metres
Varn 5°W
CHIMNEY
Small Craft Moorings
Noss Creek
Noss Plantation
Kingswear Marina
Pontoons
Peppersheen Plantation
Moored Barges
Pontoon
Rough Hole Pt
Old Mill Creek
CHIMNEY
Dart Marina
Britannia Royal Naval College
DARTMOUTH
Boat Camber
Dartmouth Yacht Club
Darthaven Marina
Waterhead Creek

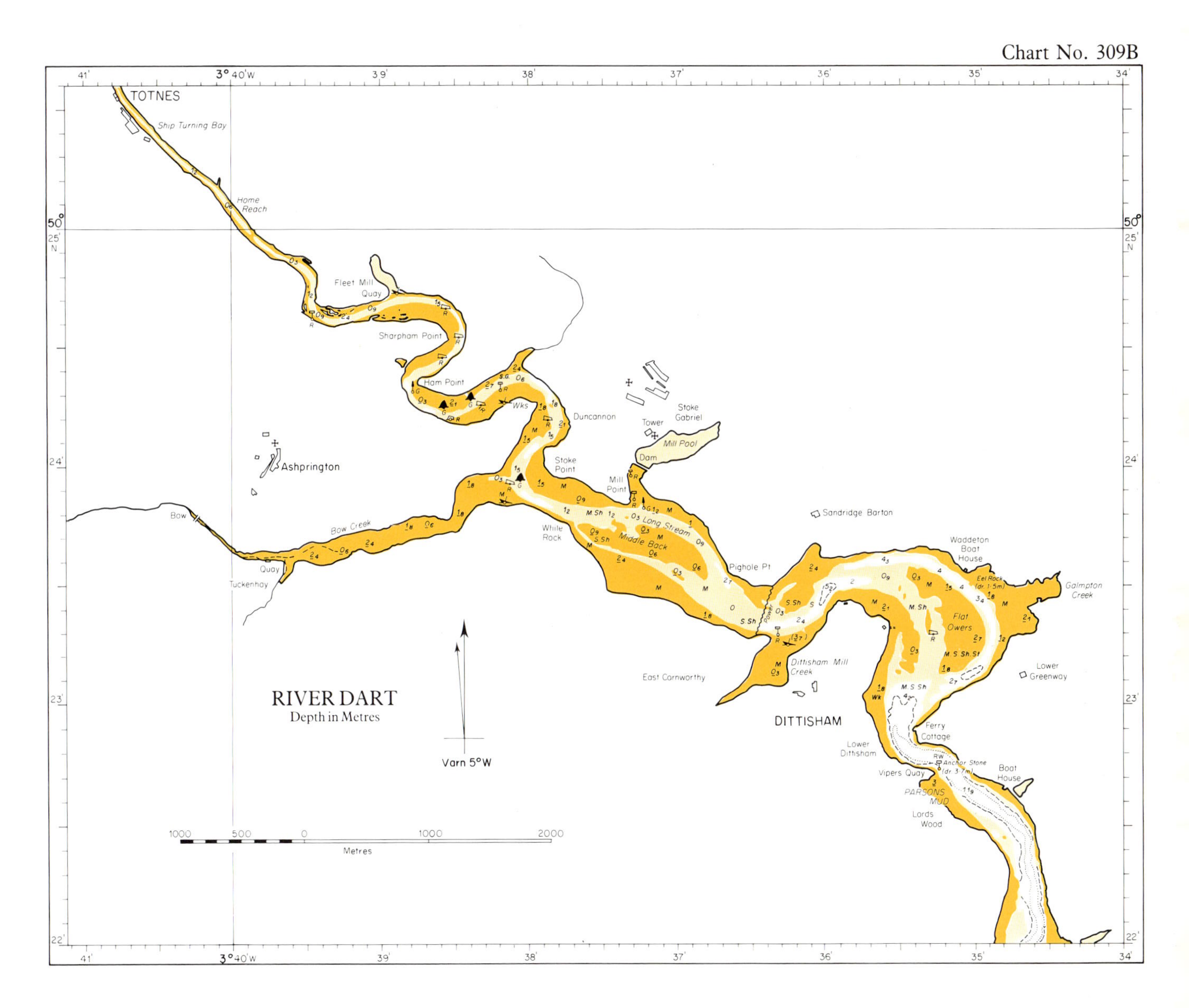
Chart No. 309B
RIVER DART
Depth in Metres
Metres
Varn 5°W
TOTNES
Ship Turning Bay
Home Reach
Fleet Mill Quay
Sharpham Point
Ham Point
Duncannon
Ashprington
Bow
Bow Creek
Quay
Tuckenhay
Stoke Point
White Rock
Mill Point
Tower
Stoke Gabriel
Mill Pool
Dam
Long Stream
Middle Back
Pighole Pt
Sandridge Barton
East Cornworthy
Dittisham Mill Creek
DITTISHAM
Waddeton Boat House
Eel Rock
Galmpton Creek
Flat Owers
Lower Greenway
Ferry Cottage
Lower Dittisham
Vipers Quay
Anchor Stone
Boat House
PARSONS MUD
Lords Wood

Salcombe

See Chart No 310.

TIDE

Time and height differences on Standard Port (for instructions in use, see p. xii)

	Lat	Long	Time Difference HW		LW		Height Difference (metres)			
PLYMOUTH See p. 28	50°22′N	4°11′W	0100 and 1300	0600 and 1800	0100 and 1300	0600 and 1800	5.5	4.4	2.2	0.8
Salcombe	50°13′N	3°47′W	0000	+0010	+0005	−0005	−0.2	−0.3	−0.1	−0.1

TIDES relative to HW Dover (approx) HW −0530. Off Salcombe in the river, the stream sets NE at −5 HW Plymouth and SW at HW +1 Plymouth.

GENERAL

Salcombe is a fine harbour devoted almost entirely to yachts. There are plenty of visitors' moorings, a visitors' pontoon and a fuel barge. For water, call the water boat by hoisting a bucket in the rigging. The only snag is the bar, see below. It is possible to anchor off in Starehole Bay or The Range.

APPROACH

There are two pairs of leading lights so, apart from the bar (see below), the entrance is simple.

CAUTIONS

1 Dangerous breaking seas may occur on the bar (about 1.2m charted depth) on onshore winds or swell, especially on the ebb or below half tide. Boats have been lost even in quite an innocent looking swell.
2 Anchoring is prohibited in the area marked.
3 Speed limit – 8 knots.

TELEPHONES

Harbour Master:	Salcombe 3791
HM Coastguard:	Chivelstone 259 (Prawle Pt) or Brixham 2156
HM Customs:	Salcombe 2835 (Local Office), or to clear call Freephone Customs Yachts or (0752) 669811
Medical:	Salcombe 2284 (Health Centre, 24 hrs)
Fuel Barge:	Salcombe 3218

VHF

Harbour Master:	Ch 16; 14
Island Cruising Club:	Ch M (base or Egremont)
Fuel barge:	Ch 6

YACHT CLUBS

Salcombe YC:	Salcombe 2593
Island Cruising Club:	Salcombe 3481

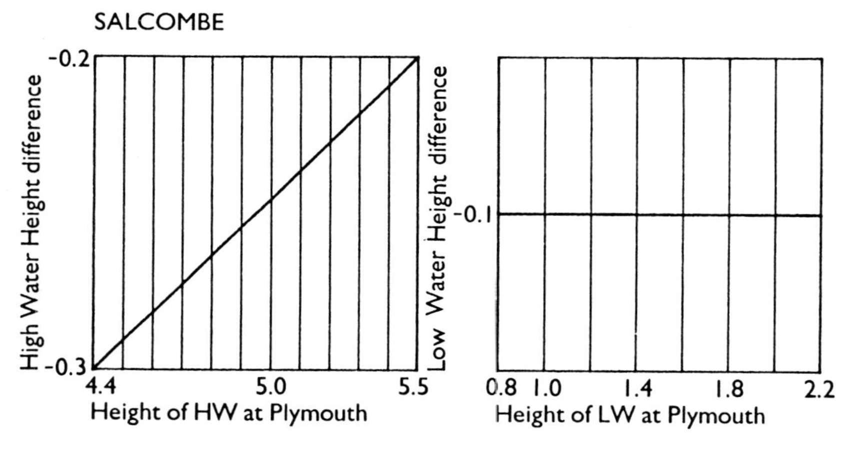

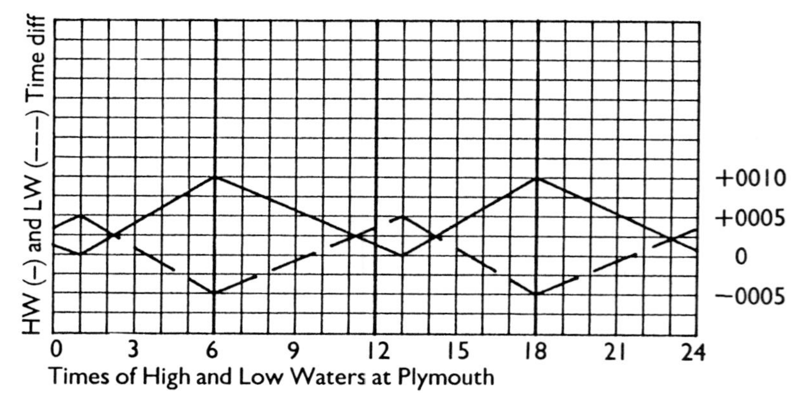

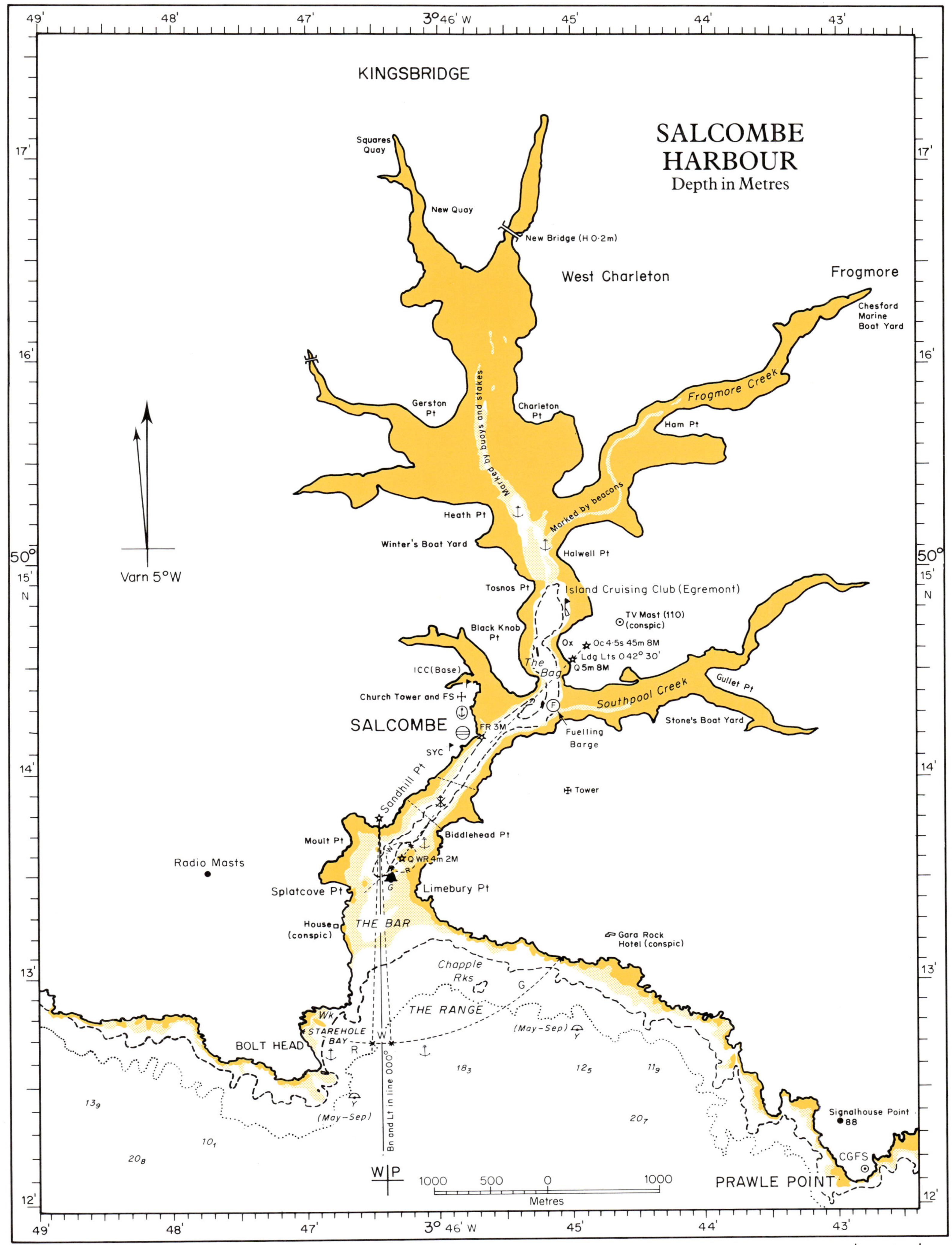

SALCOMBE HARBOUR
Depth in Metres
KINGSBRIDGE
Squares Quay
New Quay
New Bridge (H 0·2m)
West Charleton
Frogmore
Chesford Marine Boat Yard
Frogmore Creek
Ham Pt
Gerston Pt
Charleton Pt
Marked by buoys and stakes
Marked by beacons
Heath Pt
Winter's Boat Yard
Halwell Pt
Varn 5°W
Tosnos Pt
Island Cruising Club (Egremont)
TV Mast (110) (conspic)
Black Knob Pt
Oc 4·5s 45m 8M
Ldg Lts 042° 30'
Q 5m 8M
The Bag
ICC (Base)
Gullet Pt
Church Tower and FS
Southpool Creek
SALCOMBE
FR 3M
Fuelling Barge
Stone's Boat Yard
SYC
Sandhill Pt
Tower
Moult Pt
Biddlehead Pt
Radio Masts
Q WR 4m 2M
Splatcove Pt
Limebury Pt
House (conspic)
THE BAR
Gara Rock Hotel (conspic)
Chapple Rks
THE RANGE
STAREHOLE BAY
BOLT HEAD
(May–Sep)
Bn and Lt in line 000°
Signalhouse Point
88
CGFS
PRAWLE POINT
W|P
1000 500 0 1000
Metres
Way Pt: 50°12'·1N 3°46'·4W

Yealm River

See Chart No 311.

TIDE

Time and height differences on Standard Port (for instructions in use, see p. xii)

	Lat	Long	Time Difference HW		LW		Height Difference (metres)			
			0000 and 1200	0600 and 1800	0000 and 1200	0600 and 1800				
PLYMOUTH See p. 28	50°22′N	4°11′W					5.5	4.4	2.2	0.8
R Yealm Entr	50°18′N	4°04′W	+0006	+0006	+0002	+0002	−0.1	−0.1	−0.1	−0.1

TIDES relative to HW Dover (approx) HW −0540. Stream in river ebbs −0525, floods +0100.

GENERAL

The mouth of the river Yealm is within Wembury Bay which provides an anchorage in offshore winds while waiting to enter; but see below. In other than onshore winds the access is easy and, given sufficient water over the bar, safe. Shelter within is good; there is an anchorage in Cellar Bay, just over the bar, visitors' moorings and most facilities for yachts.

APPROACH

There are shoal patches and rocks encumbering Wembury Bay but an approach of 023° to St Weburgh's Church clears all dangers. There are triangular BW beacons on 088° leading to the mouth, also marked with a R can buoy (summer only). The second set of beacons which may have been familiar (1987) have been removed and a new beacon erected on the S side of the entrance.

CAUTIONS

1 The whole area is to be avoided in strong SW winds, which create a breaking sea over the shoal patches in Wembury Bay and over the bar.
2 The bar extends nearly the whole way across the entrance to the Yealm from the N.
3 There is a firing range at Wembury Point. Firing is indicated by R flags and information can be obtained from HMS Cambridge on VHF Chs 16; 12; 14 or by telephone (0752) 553740.

TELEPHONES

Harbour Master:	Plymouth 872533
HM Coastguard:	Plymouth 872301
MRSC:	Brixham 58292
HM Customs:	Freephone Customs Yachts or Plymouth 669811
Medical:	Plymouth 880392 (Doctor)

YACHT CLUB

Yealm YC:	Plymouth 872291

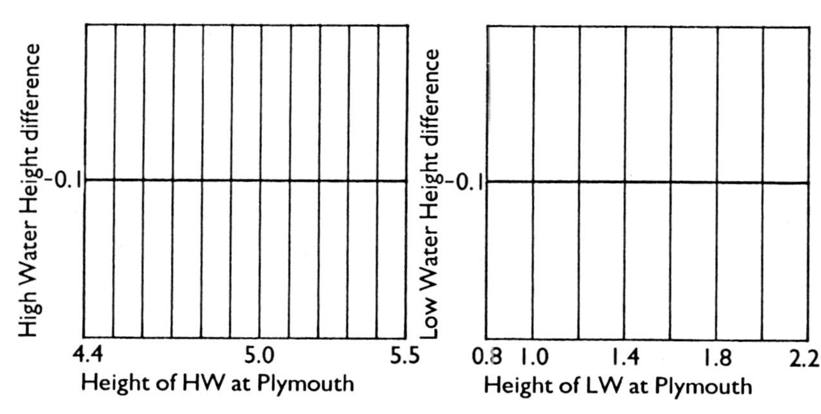

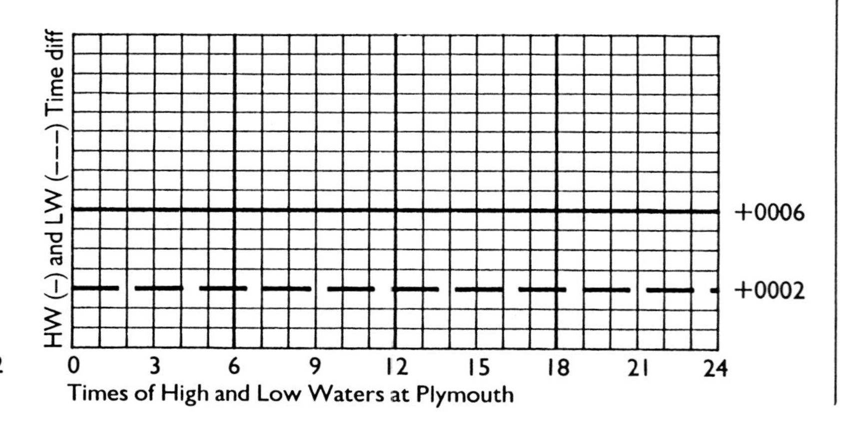

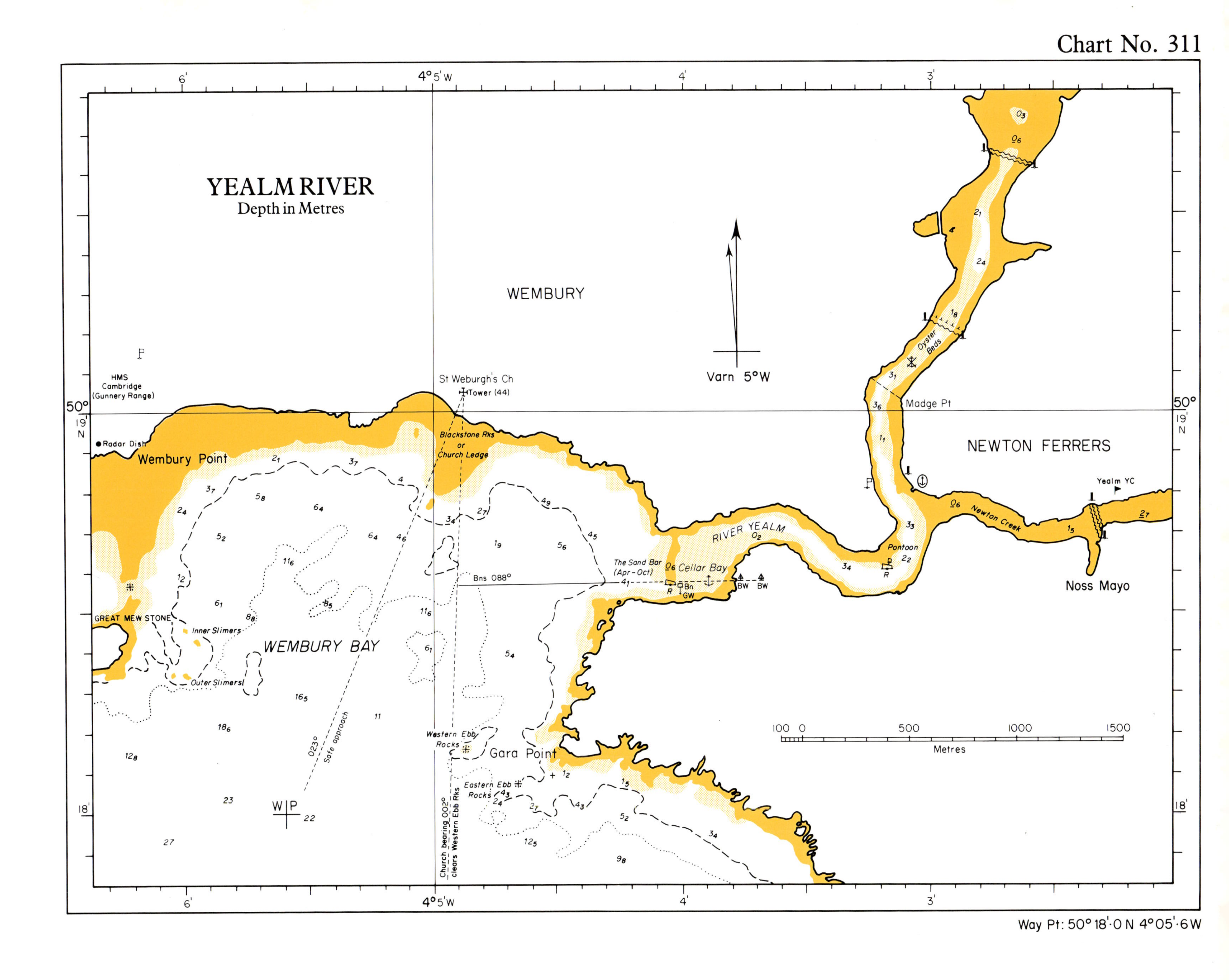
YEALM RIVER
Depth in Metres
WEMBURY
NEWTON FERRERS
Noss Mayo
Yealm YC
Newton Creek
RIVER YEALM
Cellar Bay
The Sand Bar (Apr-Oct)
Pontoon
Oyster Beds
Madge Pt
Varn 5°W
St Weburgh's Ch
Tower (44)
Blackstone Rks or Church Ledge
HMS Cambridge (Gunnery Range)
Radar Dish
Wembury Point
GREAT MEW STONE
Inner Slimers
Outer Slimers
WEMBURY BAY
Bns 088°
023° Safe approach
Western Ebb Rocks
Eastern Ebb Rocks
Gara Point
Church bearing 002° clears Western Ebb Rks
W|P
Metres
100 0 500 1000 1500
4°5'W
6'
4'
3'
50° 19' N
18'
Way Pt: 50°18'·0 N 4°05'·6 W

Approaches to Plymouth

See Chart No 312.

TIDE

See Devonport on page 28.

GENERAL

Plymouth Sound gives access to Plymouth, Devonport, Hamoaze and the Lynher River and the Tamar River, all of which are described and charted in the following pages.

APPROACH

The channel to the W of the breakwater is lit but the E channel may also be used without difficulty.

CAUTIONS

1 Large vessels use the port and should be given a wide berth.
2 There are bathing and diving areas N and S of Fort Bovisand.
3 There is a firing range at Wembury Point – see p. 22.

TELEPHONES

Queen's Harbour Master:	Plymouth 663225
Harbour Master, Cattewater:	Plymouth 665934
Harbour Master, Mill Bay Docks:	Plymouth 662191
HM Coastguard:	Plymouth 822239
MRSC:	Brixham 58292
HM Customs:	Freephone Customs Yachts or Plymouth 669811
Medical:	Plymouth 53533 (Doctor) or 668080 (Hospital)

VHF

Long Room Port Control: Ch 16; 8; 12; 14
and see succeeding pages.

YACHT CLUBS

R Western YC of England:	Plymouth 660077
R Plymouth Corinthian YC:	Plymouth 664327
Mayflower SC:	Plymouth 662526
Cawsand SC:	Plymouth 822429
Saltash SC:	Saltash 2826
Torpoint Mosquito SC:	Plymouth 812508
Weir Quay SC:	Bere Alston 840400

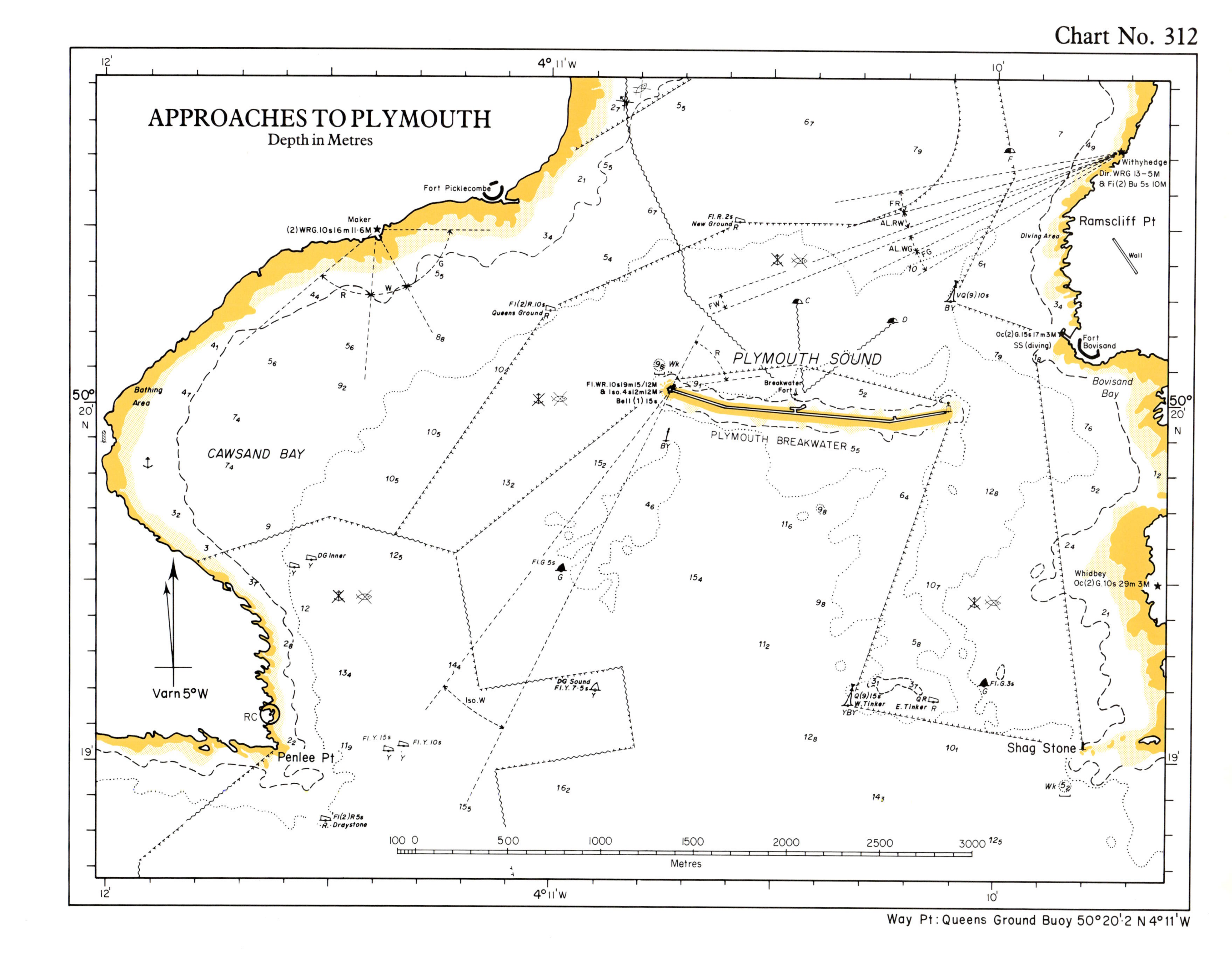
Chart No. 312
APPROACHES TO PLYMOUTH
Depth in Metres
PLYMOUTH SOUND
PLYMOUTH BREAKWATER
Breakwater Fort
CAWSAND BAY
Ramscliff Pt
Fort Bovisand
Bovisand Bay
Withyhedge Dir. WRG 13-5M & Fl(2) Bu 5s 10M
Oc(2)G.15s 17m 3M SS (diving)
Whidbey Oc(2) G.10s 29m 3M
Shag Stone
Fort Picklecombe
Maker (2) WRG.10s 16m 11-6M
Penlee Pt
Diving Area
Bathing Area
Fl.R.2s New Ground R
Fl(2)R.10s Queens Ground R
Fl.WR.10s 19m 15/12M & Iso.4s 12m 12M Bell (1) 15s
Q(9)15s W. Tinker
QR E. Tinker R
DG Inner
DG Sound Fl.Y.7·5s
Fl(2)R5s R. Draystone
Varn 5°W
Metres
Way Pt: Queens Ground Buoy 50°20'.2 N 4°11'W

Plymouth

See Chart No 313.

TIDE

Time and height differences on Standard Port (for instructions in use, see p. xii)

	Lat	Long	Time Difference HW		Time Difference LW		Height Difference (metres)			
			0000 and 1200	0600 and 1800	0000 and 1200	0600 and 1800				
PLYMOUTH (Devonport) See p. 28	50°22′N	4°11′W					5.5	4.4	2.2	0.8
Turnchapel	50°22′N	4°07′W	0000	0000	+0010	−0015	0.0	+0.1	+0.2	+0.1
Bovisand Pier	50°20′N	4°08′W	0000	−0020	0000	−0010	−0.2	−0.1	0.0	+0.1

TIDES relative to HW Dover (approx) HW −0540.

GENERAL

A new marina at Sutton Harbour has provided an alternative to the rather grim Mill Bay Docks and given easy access to the city. All facilities.

APPROACH

See page 24.

CAUTION

1 Smeaton Pass is used by large vessels, which should be given a wide berth.

TELEPHONES

See page 24.

VHF

Long Room Port Control:	Ch 16; 8; 12; 14
Sutton Harbour Radio:	Ch 16; 12; M
Queen Anne's Battery Marina:	Ch M
Sutton Harbour Marina:	Ch M

MARINAS

Sutton Harbour:	Plymouth 664186
Queen Anne's Battery:	Plymouth 671142

YACHT CLUBS

See page 24.

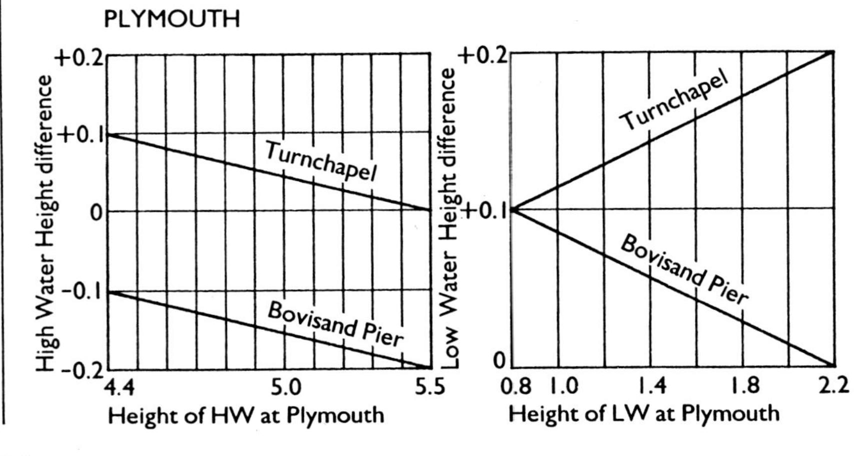

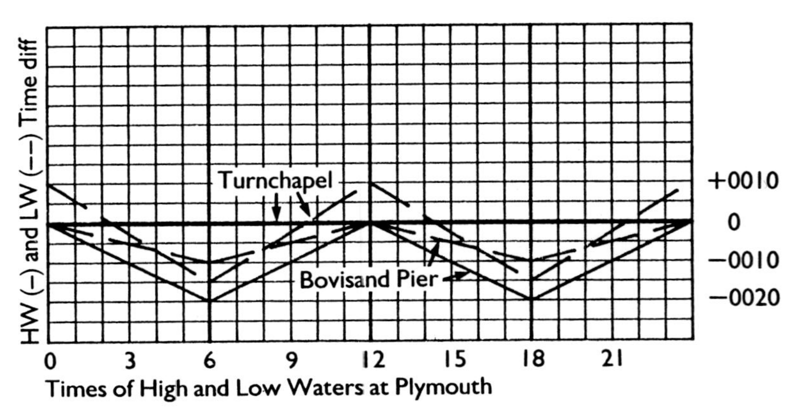

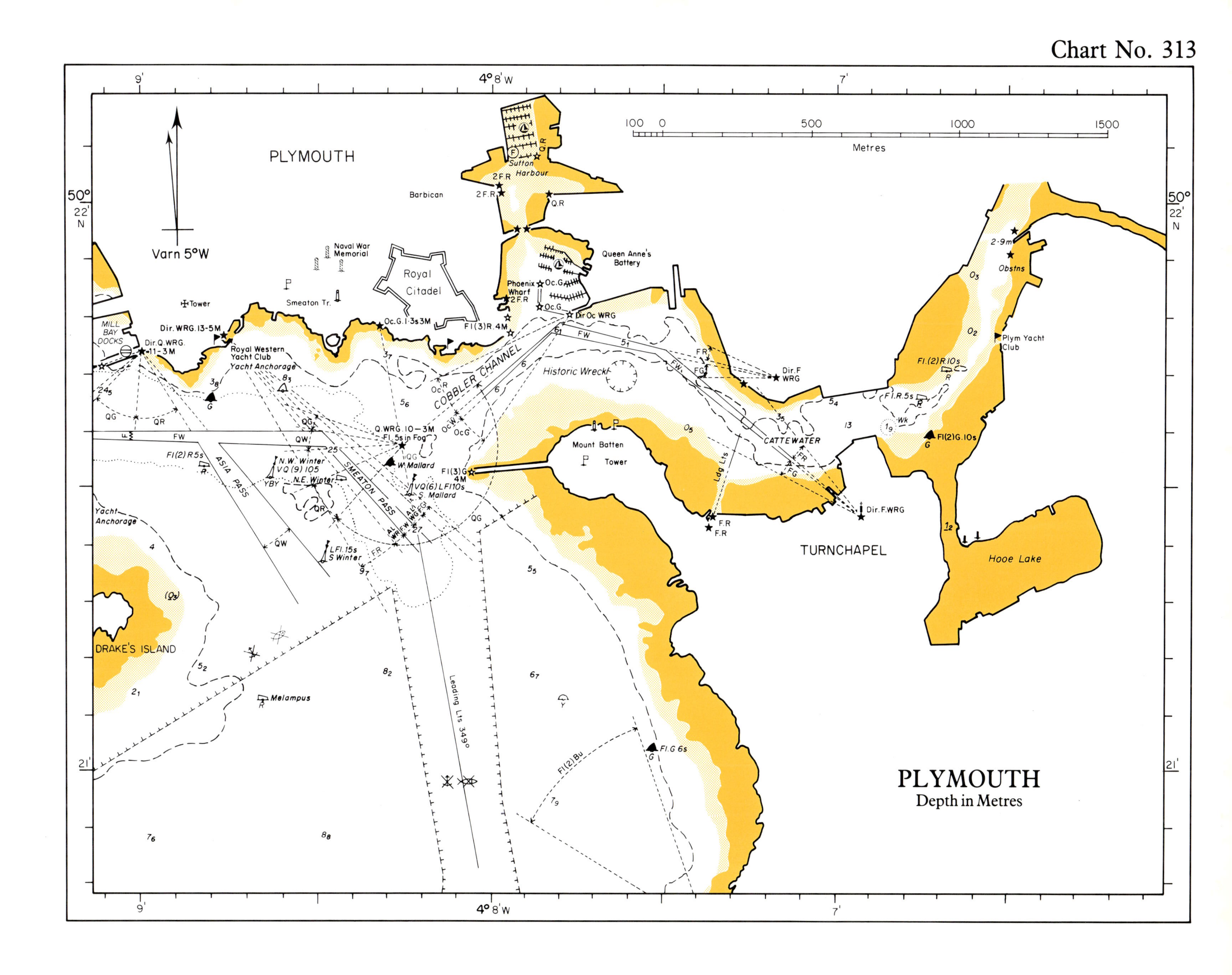

PLYMOUTH
Depth in Metres
Metres
Varn 5°W
PLYMOUTH
Barbican
Sutton Harbour
Queen Anne's Battery
Phoenix Wharf
Royal Citadel
Naval War Memorial
Smeaton Tr.
Tower
MILL BAY DOCKS
Royal Western Yacht Club
Yacht Anchorage
COBBLER CHANNEL
Historic Wreck
SMEATON PASS
ASIA PASS
W. Mallard
S. Mallard
N.W. Winter
N.E. Winter
S Winter
Yacht Anchorage
DRAKE'S ISLAND
Melampus
Leading Lts 349°
Mount Batten
Tower
CATTEWATER
TURNCHAPEL
Plym Yacht Club
Obstns
Hooe Lake

Plymouth (Devonport)

See Chart No 314.

TIDE

Plymouth (Devonport) is a Standard Port. Use the tidal curve below.

Times				Height (metres)			
HW		LW					
				MHWS	MHWN	MLWN	MLWS
0000	0600	0000	0600	5.5	4.4	2.2	0.8
1200	1800	1200	1800				

TIDES relative to HW Dover (approx) HW −0540.

GENERAL

Devonport is a naval base but there is a large marina (Mayflower) and all facilities for yachts.

APPROACH

See Approaches to Plymouth on page 24.

TELEPHONES

See page 24.

VHF

Mayflower
Marina: Ch M

Also see Plymouth on page 26.

MARINA

Mayflower
Marina: Plymouth 556633

YACHT CLUBS

See page 24.

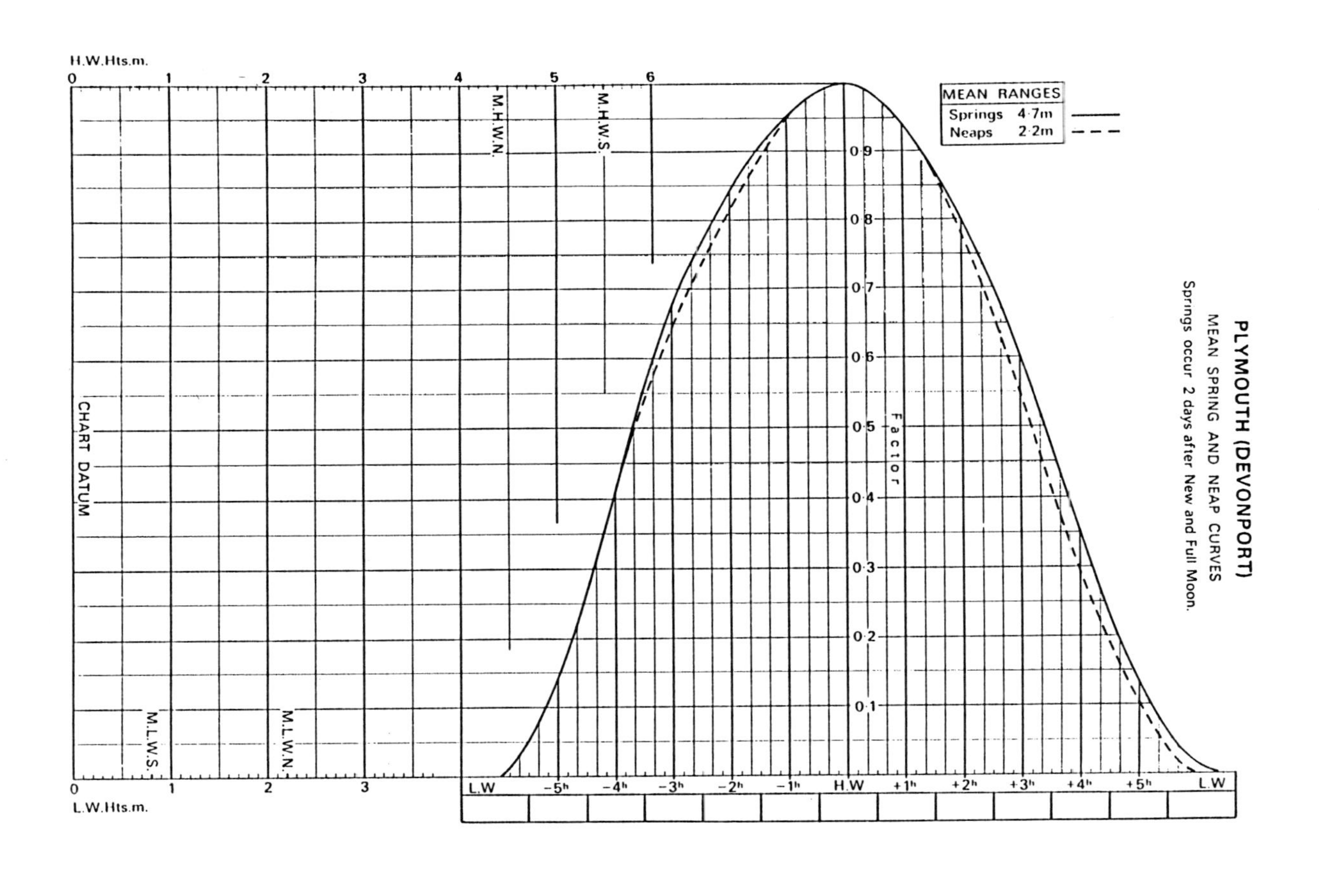

Chart No. 314

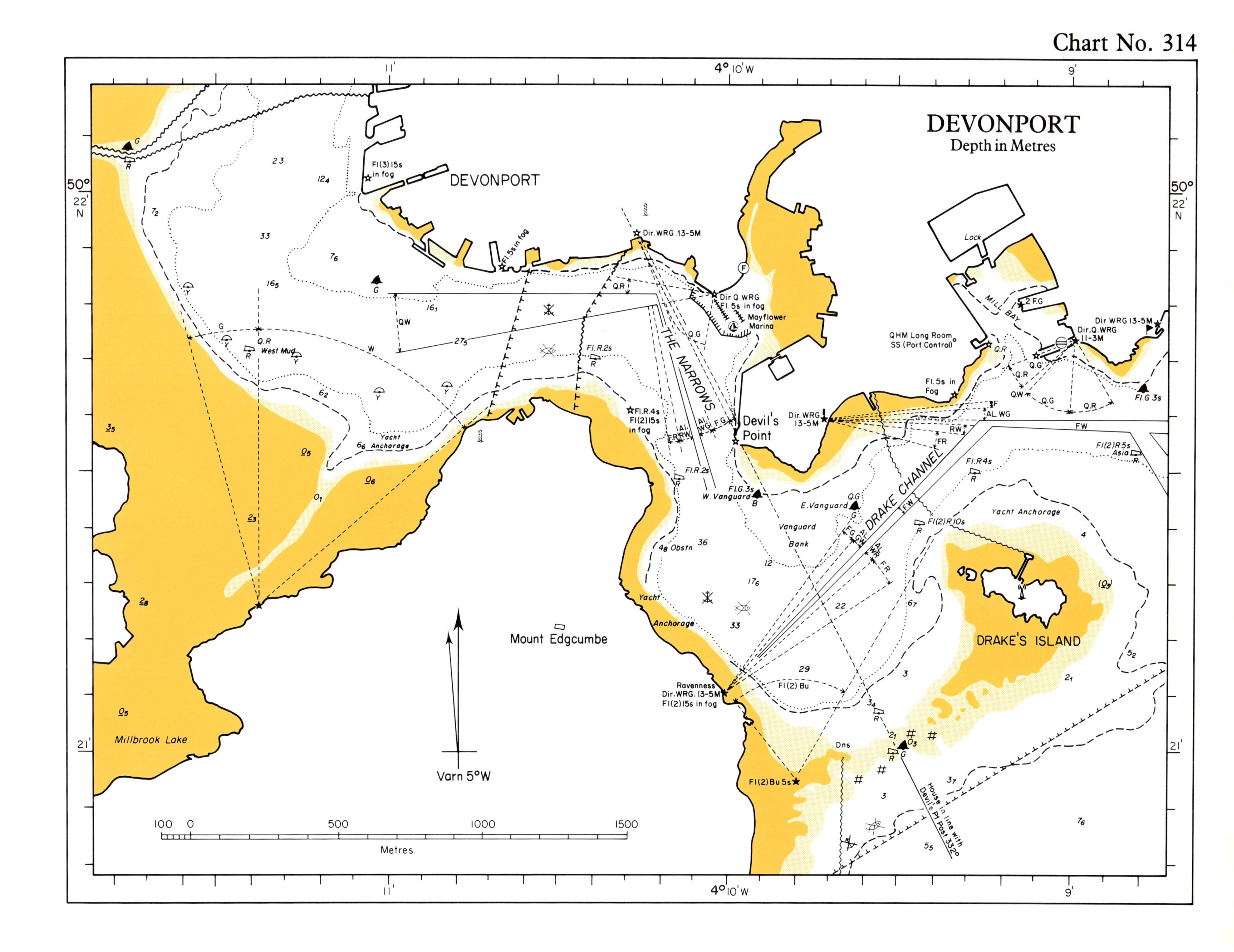
DEVONPORT
Depth in Metres
DEVONPORT
Fl(3)15s in fog
Fl.5s in fog
Dir.WRG.13·5M
Dir Q WRG
Fl.5s in fog
Mayflower Marina
THE NARROWS
Devil's Point
Dir.WRG 13·5M
Fl.5s in Fog
QHM Long Room
SS (Port Control)
MILL BAY
Lock
2 F.G
Dir WRG 13·5M
Dir.Q.WRG 11·3M
Fl.G 3s
Fl(2)R5s Asia
Fl.R4s
DRAKE CHANNEL
Fl(2)R10s
Yacht Anchorage
DRAKE'S ISLAND
House in line with Devil's Pt Post 332°
Fl.R.4s
Fl(2)15s in fog
Fl.R.2s
Fl.G.3s
W. Vanguard
E.Vanguard
Vanguard Bank
48 Obstn
Yacht Anchorage
Ravenness
Dir.WRG. 13·5M
Fl(2)15s in fog
Fl(2) Bu
Fl(2)Bu 5s
Dns
Mount Edgcumbe
Varn 5°W
Metres
West Mud
Millbrook Lake
50° 22' N
21'
11'
4°10'W
9'

Hamoaze and Lynher River

See Chart No 315.

TIDE

Time and height differences on Standard Port (for instructions in use, see p. xii)

	Lat	Long	Time Difference HW		LW		Height Difference (metres)			
PLYMOUTH See p. 28	50°22′N	4°11′W	0000 and 1200	0600 and 1800	0000 and 1200	0600 and 1800	5.5	4.4	2.2	0.8
Jupiter Point	50°23′N	4°14′W	+0010	+0005	0000	−0005	0.0	0.0	+0.1	0.0
St Germans	50°23′N	4°18′W	0000	0000	+0020	+0020	−0.3	−0.1	0.0	+0.2

GENERAL

The Lynher River provides some pleasant anchorages, as shown on the chart, as an alternative to the marinas available nearer the city.

APPROACH

The approach is simple provided that the pilot is careful to avoid the various mud banks which are clearly marked.

TELEPHONES

See page 24.

VHF

Mayflower
Marina: Ch M
Also see Plymouth on page 26.

MARINA

Mayflower
Marina: Plymouth 556633

YACHT CLUBS

See page 24.

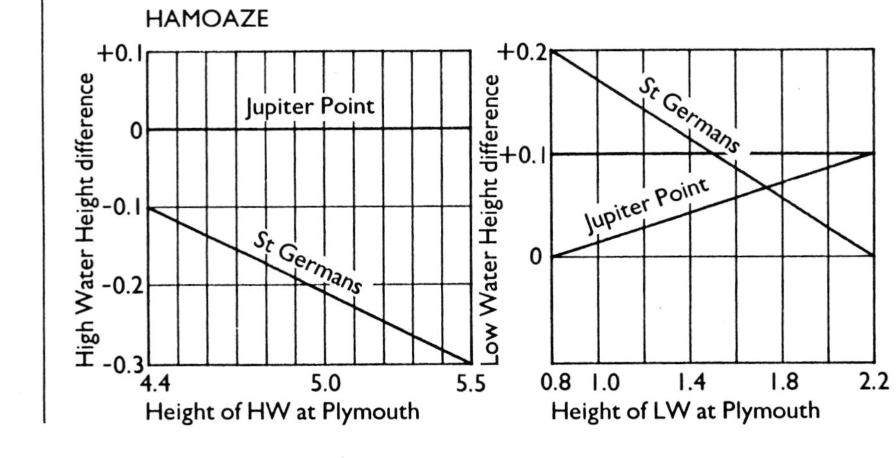

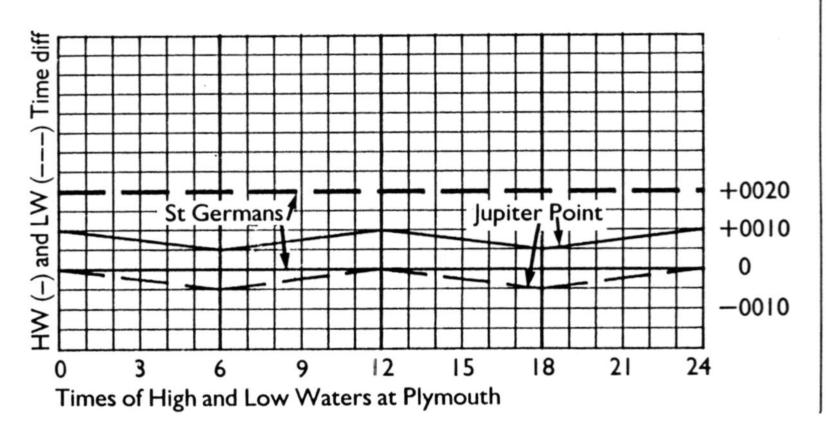

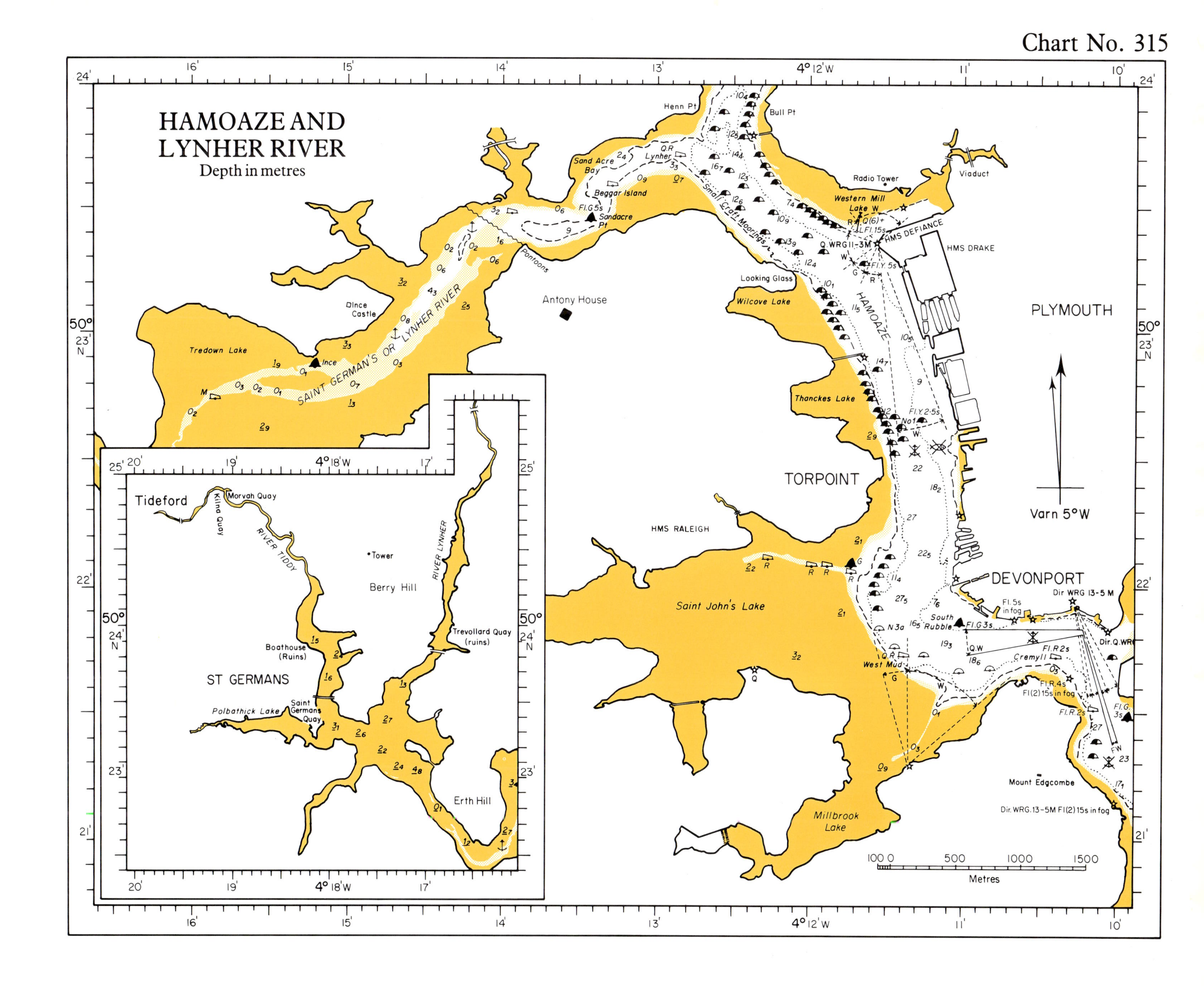
HAMOAZE AND
LYNHER RIVER
Depth in metres
PLYMOUTH
HMS DRAKE
HMS DEFIANCE
Viaduct
Radio Tower
Western Mill Lake
Bull Pt
Henn Pt
Sand Acre Bay
Beggar Island
Sandacre Pt
Lynher
Small Craft Moorings
HAMOAZE
Looking Glass
Wilcove Lake
Thanckes Lake
Pontoons
Antony House
Ince Castle
Ince
Tredown Lake
SAINT GERMAN'S OR LYNHER RIVER
TORPOINT
HMS RALEIGH
Saint John's Lake
Millbrook Lake
DEVONPORT
South Rubble
West Mud
Cremyll
Mount Edgcombe
Varn 5°W
Metres
Tideford
Morvah Quay
Kilna Quay
RIVER TIDDY
Tower
Berry Hill
RIVER LYNHER
Trevollard Quay (ruins)
Boathouse (Ruins)
ST GERMANS
Saint Germans Quay
Polbathick Lake
Erth Hill

River Tamar

See Chart No 316.

TIDE

Time and height differences on Standard Port (for instructions in use, see p. xii)

	Lat	Long	Time Difference				Height Difference (metres)			
			HW		LW					
PLYMOUTH See p. 28	50°22′N	4°11′W	0000 and 1200	0600 and 1800	0000 and 1200	0600 and 1800	5.5	4.4	2.2	0.8
Saltash	50°24′N	4°12′W	0000	+0010	0000	−0005	+0.1	+0.1	+0.1	+0.1
Cargreen	50°26′N	4°12′W	0000	+0020	+0020	+0020	0.0	0.0	−0.1	0.0
Cotehele Quay	50°29′N	4°13′W	0000	+0020	+0045	+0045	−0.9	−0.9	−0.8	−0.4

GENERAL

The River Tamar provides a pleasant cruising area, though it shallows quite quickly and requires care to keep off the mud. Landing is possible at Cargreen, Weirquay and a number of other small quays until Cotehele and Calstock are reached. It is even possible to float further up if the bridge at Calstock can be negotiated. There are yacht yards providing most facilities.

APPROACH

See page 30.

TELEPHONES

Queen's Harbour Master:	Plymouth 663225
Harbour Master, Cattewater:	Plymouth 665934
Harbour Master, Mill Bay Docks:	Plymouth 662191
HM Coastguard:	Plymouth 822239
MRSC:	Brixham 58292
HM Customs:	Freephone Customs Yachts or Plymouth 669811
Medical:	Plymouth 53533 (Doctor) or 668080 (Hospital)

YACHT CLUBS

Weir Quay YC:	Bere Alston 840400
R Western YC of England:	Plymouth 660077
R Plymouth Corinthian YC:	Plymouth 664327
Mayflower SC:	Plymouth 662526
Cawsand SC:	Plymouth 822429
Saltash SC:	Saltash 2826
Torpoint Mosquito SC:	Plymouth 812508
Weir Quay SC:	Bere Alston 840400

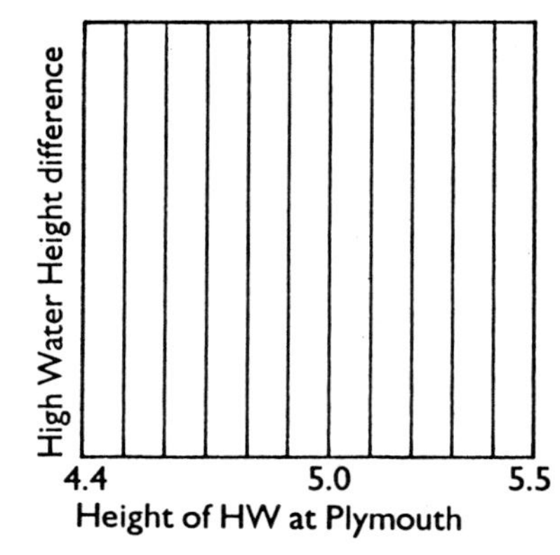

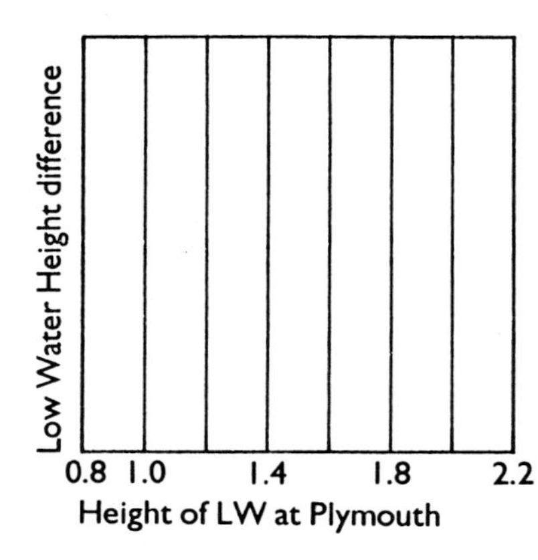

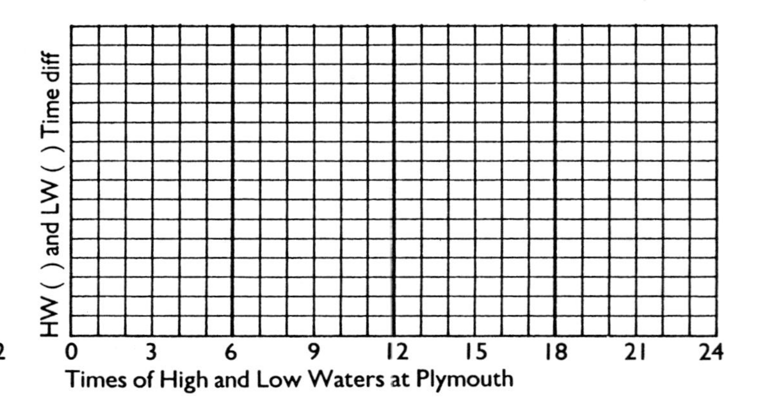

Chart No. 316a

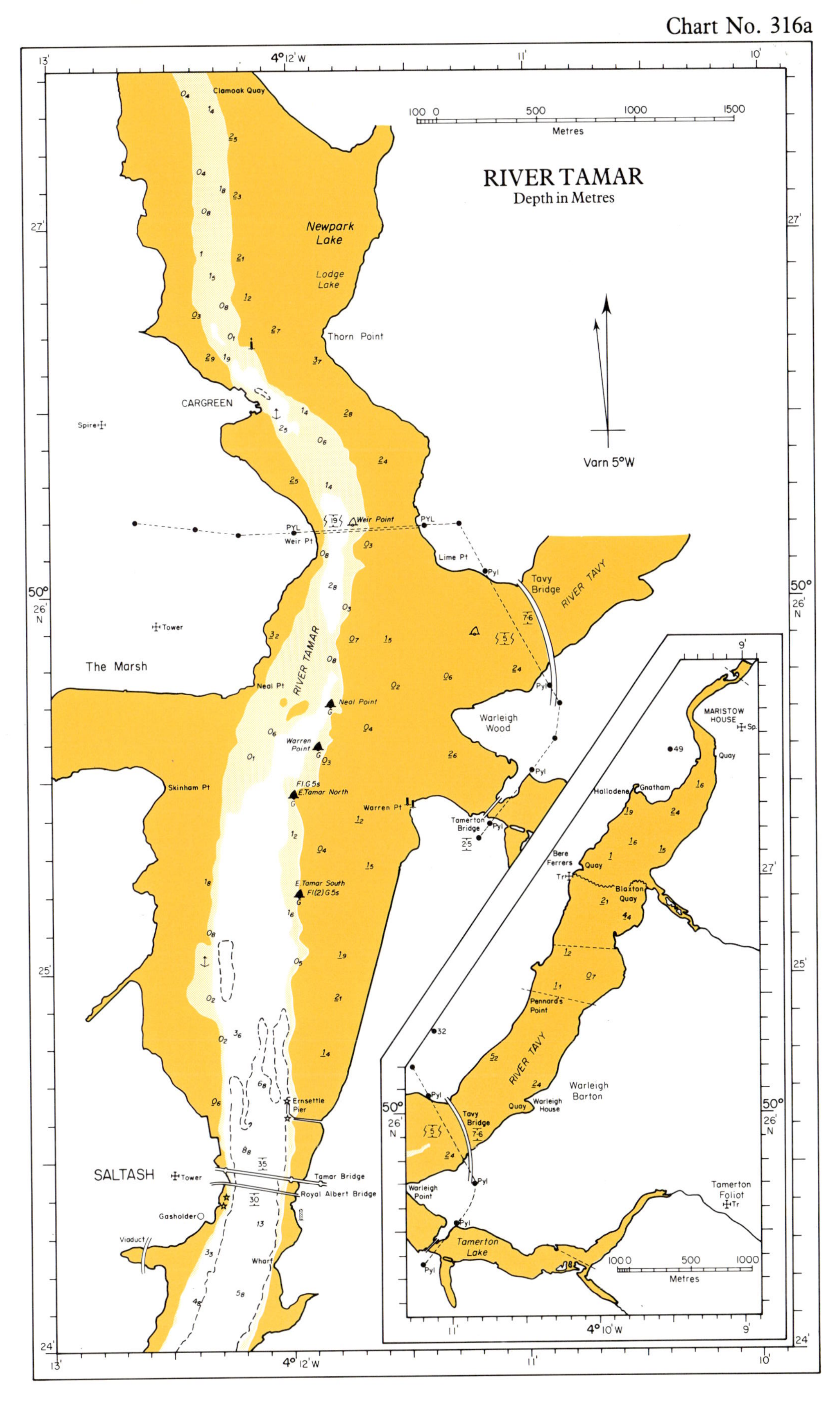

Chart No. 316b

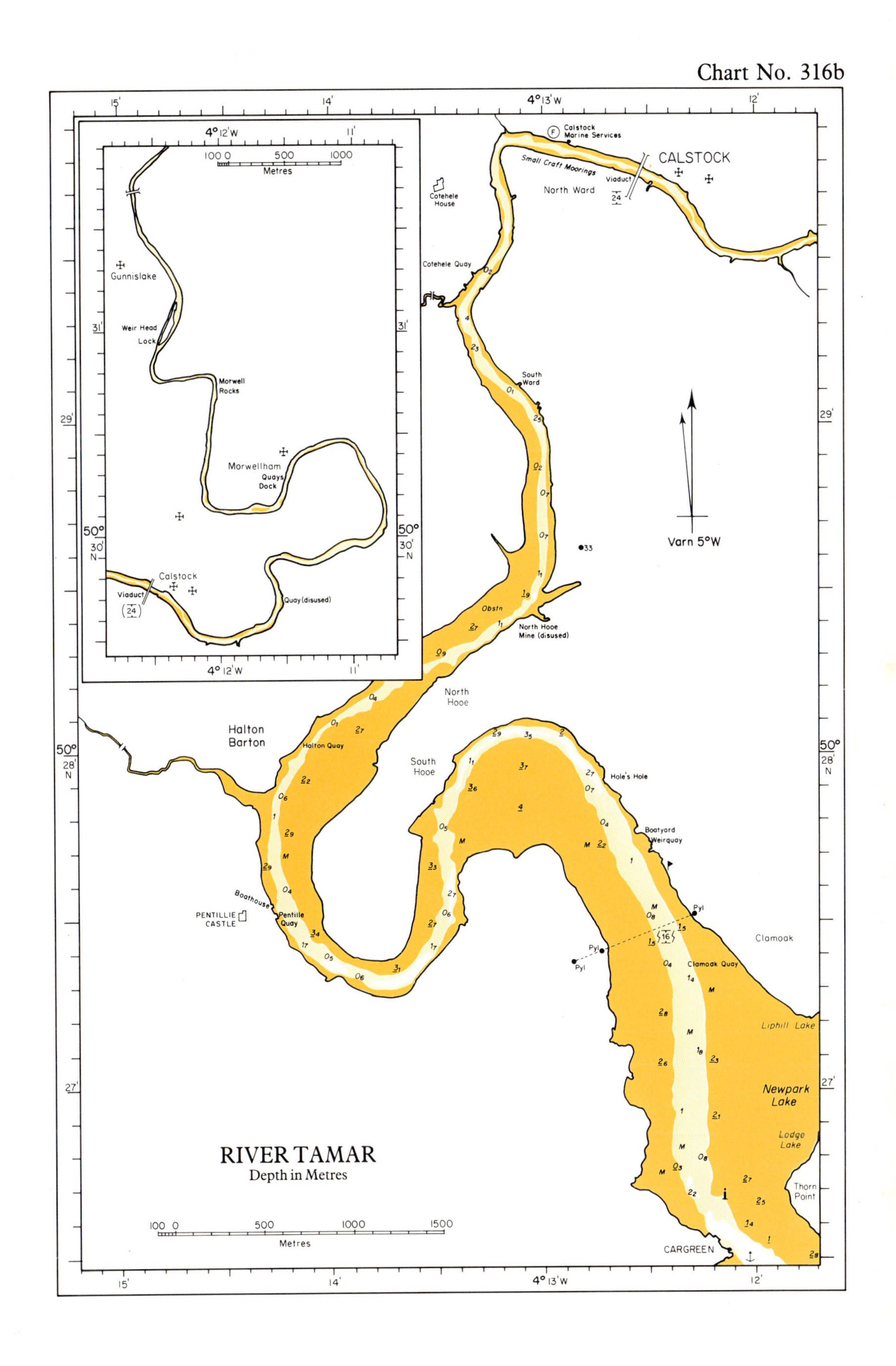

Looe

See Chart No 317.

TIDE

Time and height differences on Standard Port (for instructions in use, see p. xii)

	Lat	Long	Time Difference HW		Time Difference LW		Height Difference (metres)			
PLYMOUTH See p. 28	50°22′N	4°11′W	0000 and 1200	0600 and 1800	0000 and 1200	0600 and 1800	5.5	4.4	2.2	0.8
Looe	50°21′N	4°27′W	−0010	−0010	−0005	−0005	−0.1	−0.2	−0.2	−0.2

TIDES relative to HW Dover (approx) HW −0545. Stream sets W −0200; E +0300.

GENERAL

Looe is a holiday resort with a drying harbour. Access is good in suitable weather, as is the anchorage off.

APPROACH

The safest approach is to keep St Mary's Church (East Looe) bearing about 290° or, at night, keep within the W sector of the light. Yachts should anchor off or go alongside the W wall above the cables and pipelines near St Nicholas Church (West Looe) where there is between 2.6m and 3.1m at MHWS. All facilities.

CAUTIONS

1 Do not attempt to enter in strong SE winds which produce a breaking sea.
2 Note the remarks on Approach above.

TELEPHONES

Harbour Master:	Looe 2839
HM Coastguard:	Looe 2138
MRSC:	Brixham 58292
HM Customs:	Freephone Customs Yachts or (0752) 669811
Medical:	Looe 3195 (Doctor)

YACHT CLUB

Looe SC: Looe 2559

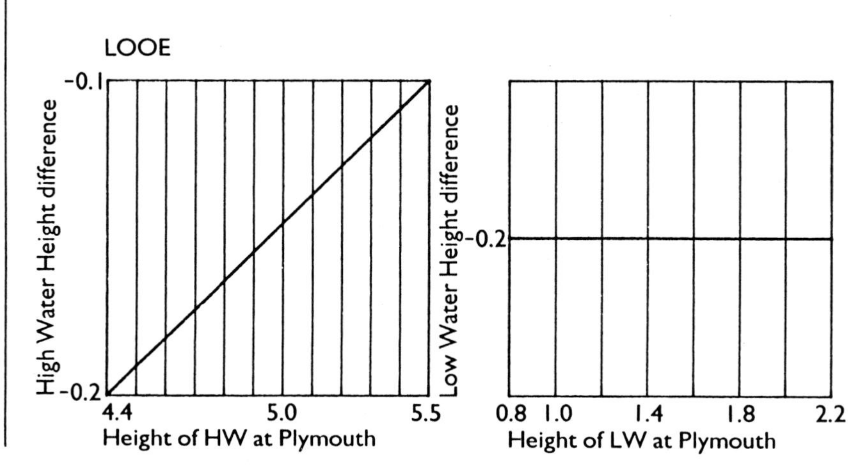

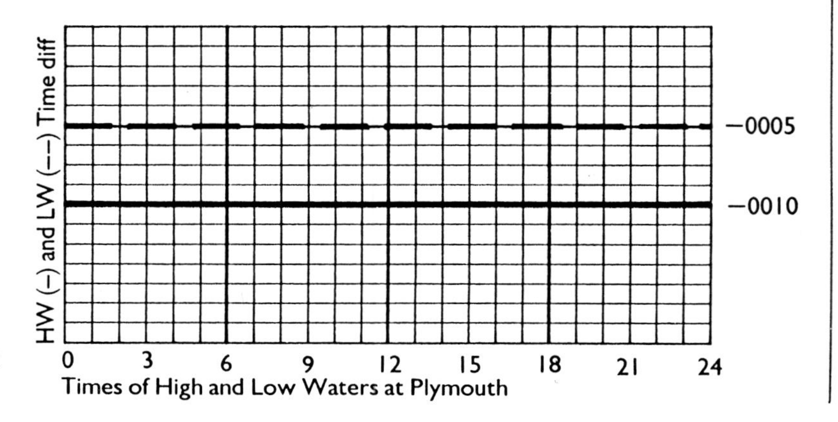

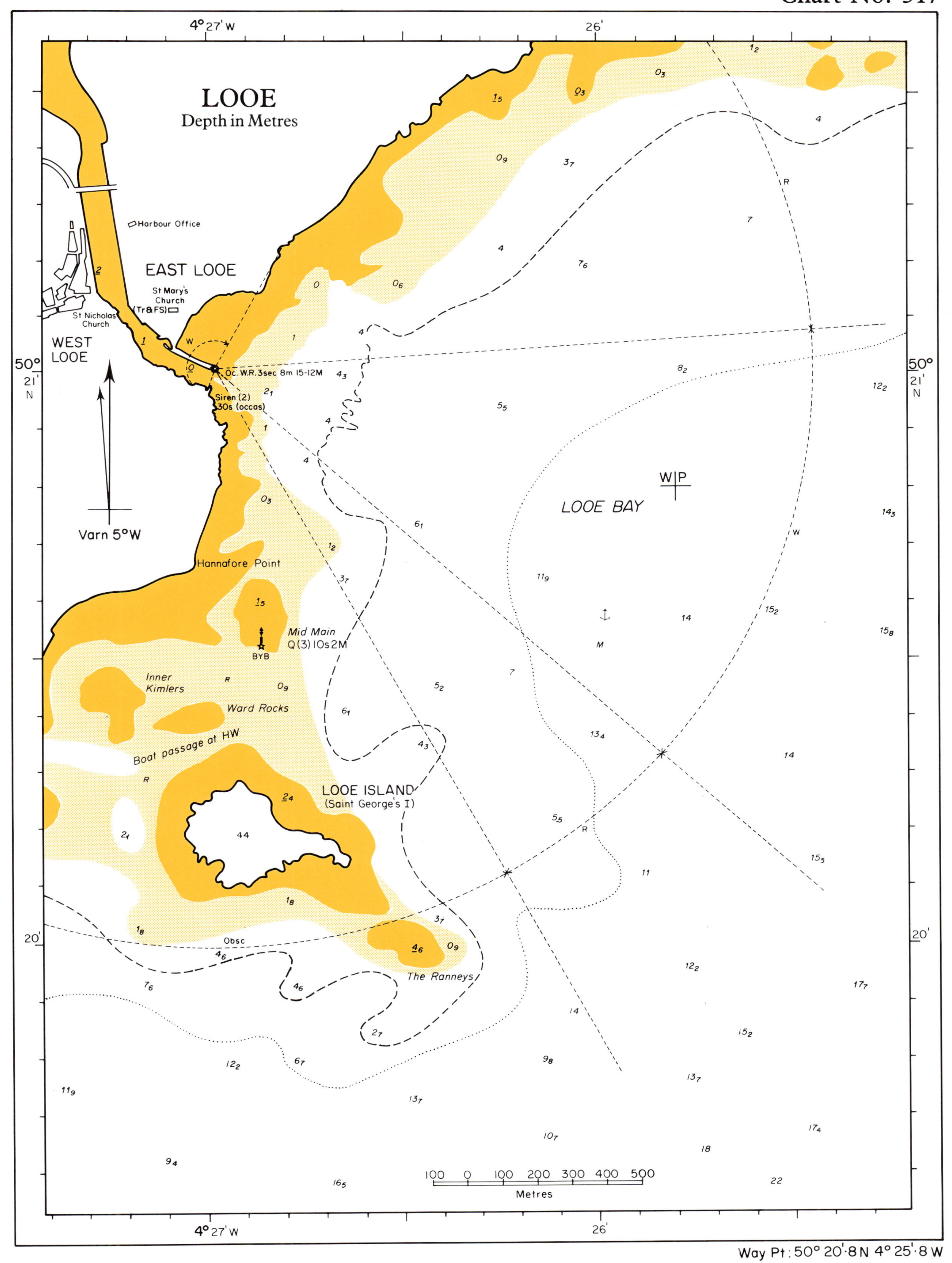
LOOE
Depth in Metres
4° 27' W
26'
Harbour Office
EAST LOOE
St Mary's Church (Tr & FS)
St Nicholas Church
WEST LOOE
Oc. W.R. 3sec 8m 15-12M
Siren (2) 30s (occas)
Varn 5°W
Hannafore Point
Mid Main Q(3) 10s 2M
BYB
Inner Kimlers
Ward Rocks
Boat passage at HW
LOOE ISLAND (Saint George's I)
The Ranneys
Obsc
LOOE BAY
WP
50° 21' N
20'
Metres
100 0 100 200 300 400 500

Way Pt: 50° 20'·8 N 4° 25'·8 W

Polperro

See Chart No 318.

TIDE

TIDES – see page 34.

TIDES relative to HW Dover (approx) HW −0550

GENERAL

Polperro is a tiny fishing and tourist harbour with room only for a few craft. The harbour dries and has about 1.5m at half tide. In bad weather the harbour entrance is closed (R lt or B ball), though remains drying. Few facilities.

APPROACH

Note the Polca rocks in the entrance. There are mooring buoys off and space may be found against the E wall just inside the entrance. The holding ground off is not good.

CAUTION

1 Do not attempt to enter at night or in bad weather.

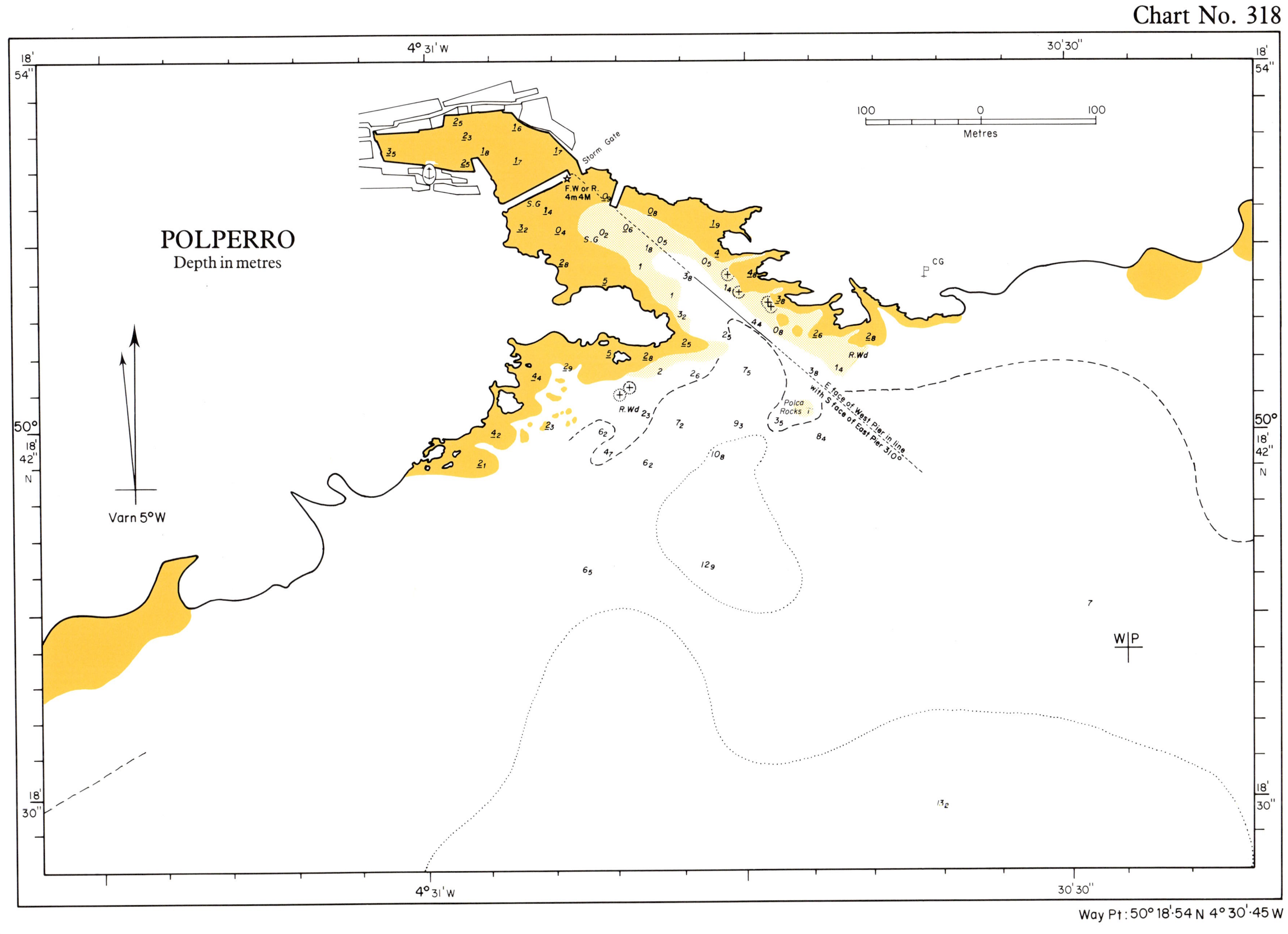

Way Pt: 50°18'·54 N 4°30'·45 W

Fowey

See Chart No 319.

TIDE

Time and height differences on Standard Port (for instructions in use, see p. xii)

	Lat	Long	Time Difference HW		Time Difference LW		Height Difference (metres)			
PLYMOUTH See p. 28	50°22′N	4°11′W	0000 and 1200	0600 and 1800	0000 and 1200	0600 and 1800	5.5	4.4	2.2	0.8
Fowey	50°20′N	4°38′W	−0010	−0015	−0010	−0005	−0.1	−0.1	−0.2	−0.2
Lostwithiel	50°24′N	4°40′W	+0005	−0010	dries		−4.1	−4.1	dries	

TIDES relative to HW Dover (approx) HW −0555. Stream sets in +0030; out −0555.

GENERAL

Fowey is a delightful town with a deep water harbour and a safe approach. All facilities.

APPROACH

The W sectors of the two lts avoid all dangers. There are visitors' moorings and pontoons on the E side about half a mile inside the entrance. Strong onshore winds may cause a heavy swell which may be avoided by moving further up the harbour.

CAUTIONS

1 There is a speed limit of 6 knots.
2 There is considerable commercial traffic.
3 Care must be taken if anchoring. Keep clear of the fairway and buoy the anchor.

TELEPHONES

Harbour Master:	Fowey 2471
HM Coastguard:	Polruan 228
MRSC:	Brixham 58292
HM Customs:	Freephone Customs Yachts or (0752) 669811
Medical:	Fowey 2451 (Doctor)

VHF

Fowey Harbour:	Ch 16; 11; 12
Polruan Ferry (water taxi):	Ch 6

YACHT CLUBS

R Fowey YC:	Fowey 2245
Fowey Gallants SC:	Fowey 2335

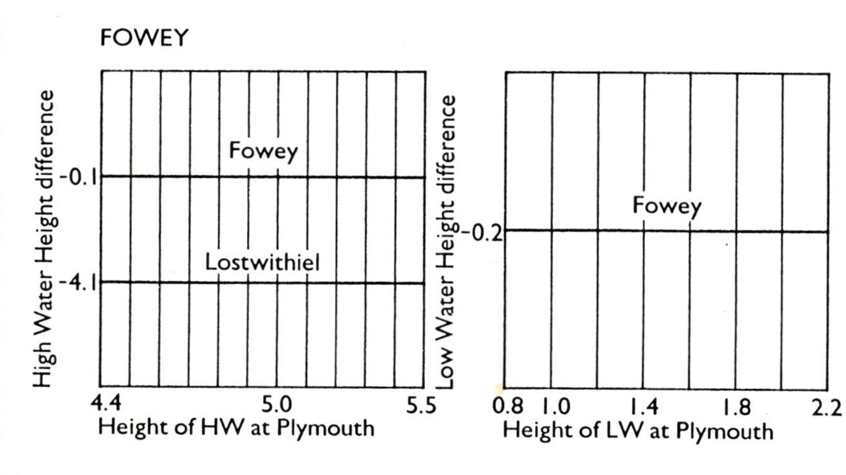

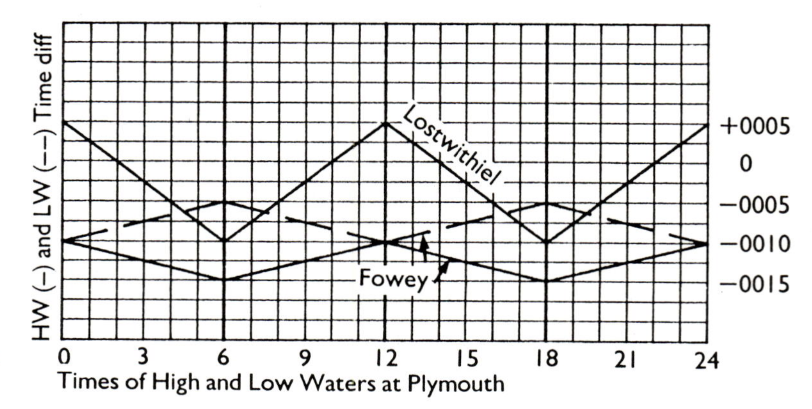

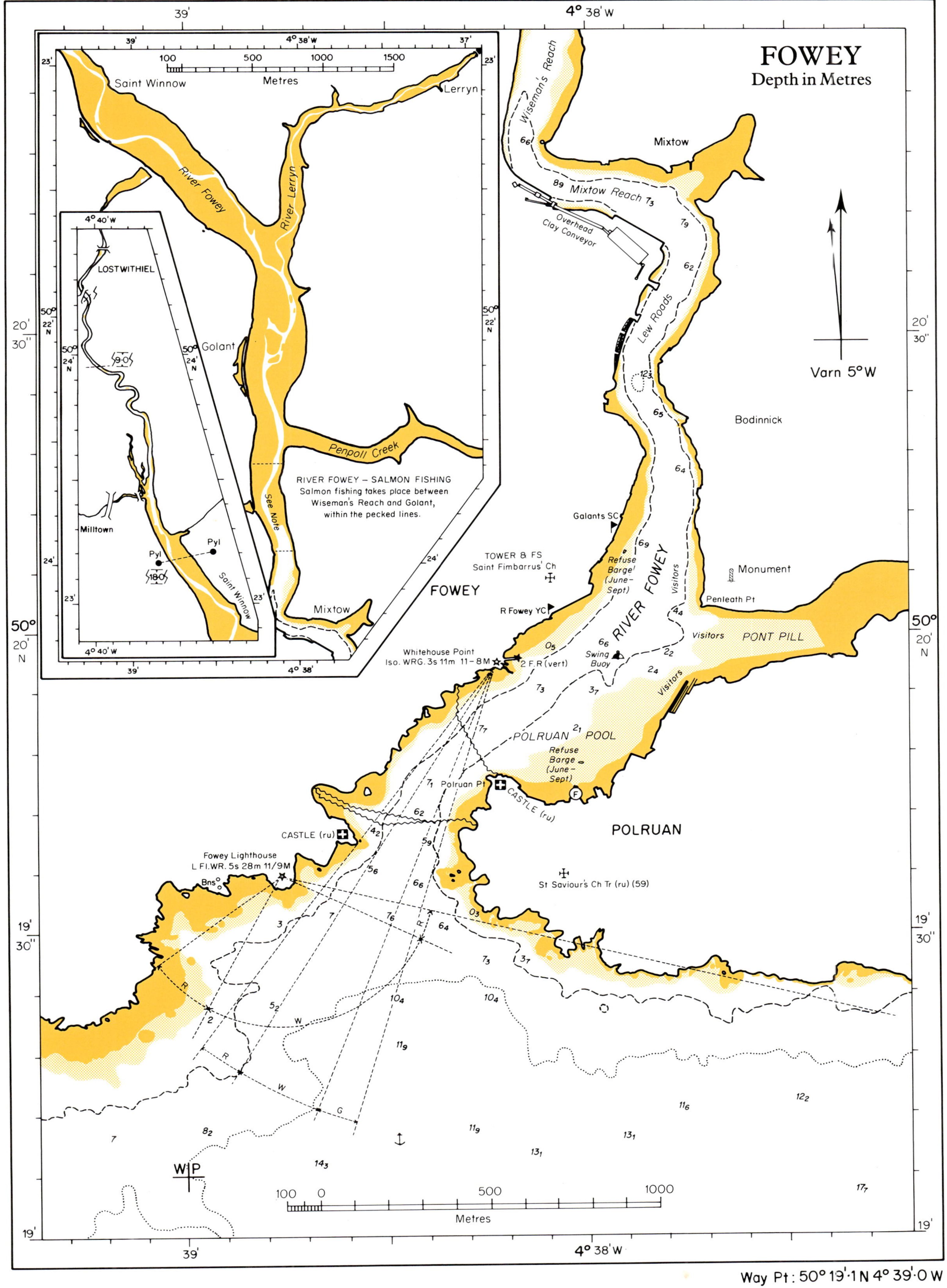
FOWEY
Depth in Metres
Varn 5°W
Mixtow
Wiseman's Reach
Mixtow Reach
Overhead Clay Conveyor
Lew Roads
Bodinnick
Galants SC
Refuse Barge (June–Sept)
RIVER FOWEY
Visitors
Monument
Penleath Pt
PONT PILL
TOWER & FS
Saint Fimbarrus' Ch
FOWEY
R Fowey YC
Whitehouse Point
Iso. WRG. 3s 11m 11–8M
2 F.R (vert)
Swing Buoy
POLRUAN POOL
Polruan Pt
CASTLE (ru)
POLRUAN
St Saviour's Ch Tr (ru) (59)
Fowey Lighthouse
L Fl.WR. 5s 28m 11/9M
Bns
WP
Metres
Saint Winnow
River Fowey
River Lerryn
Lerryn
Golant
Penpoll Creek
See Note
RIVER FOWEY – SALMON FISHING
Salmon fishing takes place between Wiseman's Reach and Golant, within the pecked lines.
LOSTWITHIEL
Milltown
Pyl
Saint Winnow

Way Pt: 50° 19'·1 N 4° 39'·0 W

Par and Charlestown

See Chart Nos 320A and 320B.

TIDE

Time and height differences on Standard Port (for instructions in use, see p. xii)

	Lat	Long	Time Difference HW		Time Difference LW		Height Difference (metres)			
			0000 and 1200	0600 and 1800	0000 and 1200	0600 and 1800				
PLYMOUTH See p. 28	50°22′N	4°11′W					5.5	4.4	2.2	0.8
Par	50°21′N	4°42′W	−0005	−0015	0000	−0010	−0.4	−0.4	−0.4	−0.2

TIDES relative to HW Dover (approx) HW −0555. Stream (at sea) sets E +0200; W −0300.

GENERAL

Both little ports specialise in the export of china clay. Par dries and is of little interest to yachtsmen. Charlestown has a wet basin with 3.7m; a G lt or R flag – harbour open; a R lt or B shape – harbour closed. In both cases prior arrangements must be made before entering – see VHF and telephone details below.

APPROACH

Par is unlit and has Killyvarder Rock 4 cables SE of the entrance. Charlestown has a safe approach from the SE. It is possible to anchor off both ports in suitable weather.

CAUTIONS

1 Do not attempt to enter without prior notice and agreement.
2 Do not attempt to enter except in good weather.

TELEPHONES

Harbour Master (Par):	Par 2282
Harbour Master (Charlestown):	St Austell 3331

VHF

Par:	Ch 16; 12 (−2HW+1)
Charlestown:	Ch 16; 14 (−1HW+1)

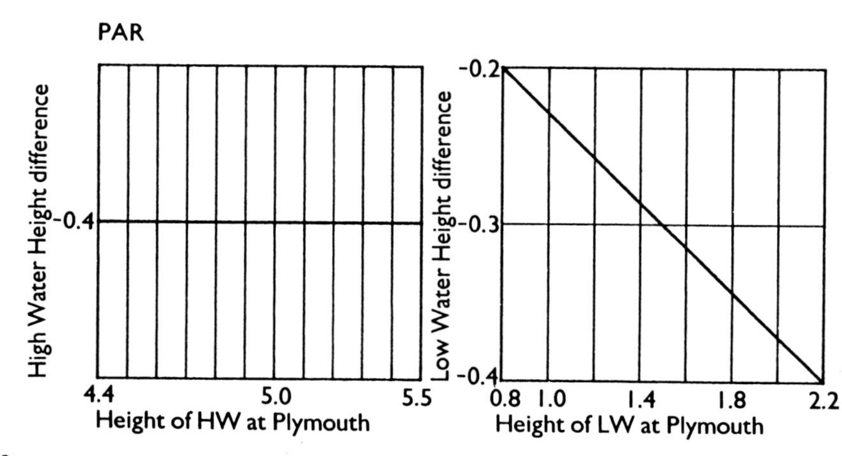

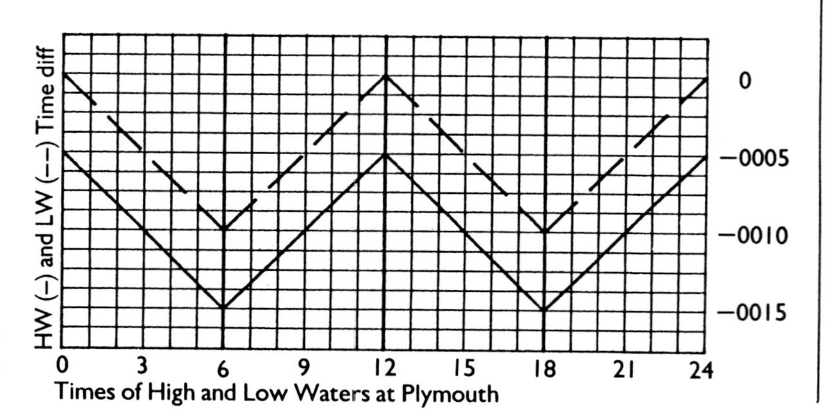

Chart No. 320a

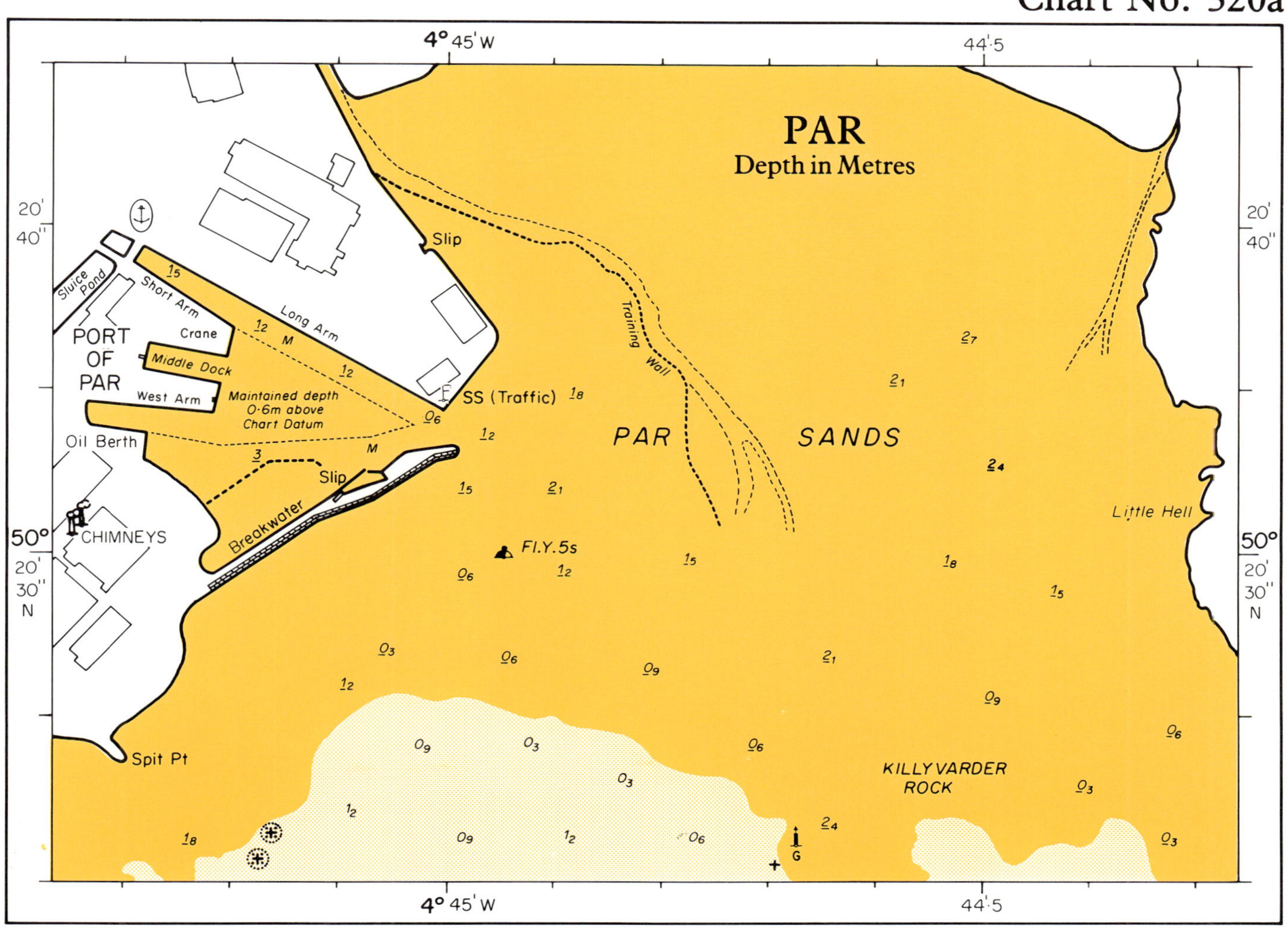

Chart No. 320b

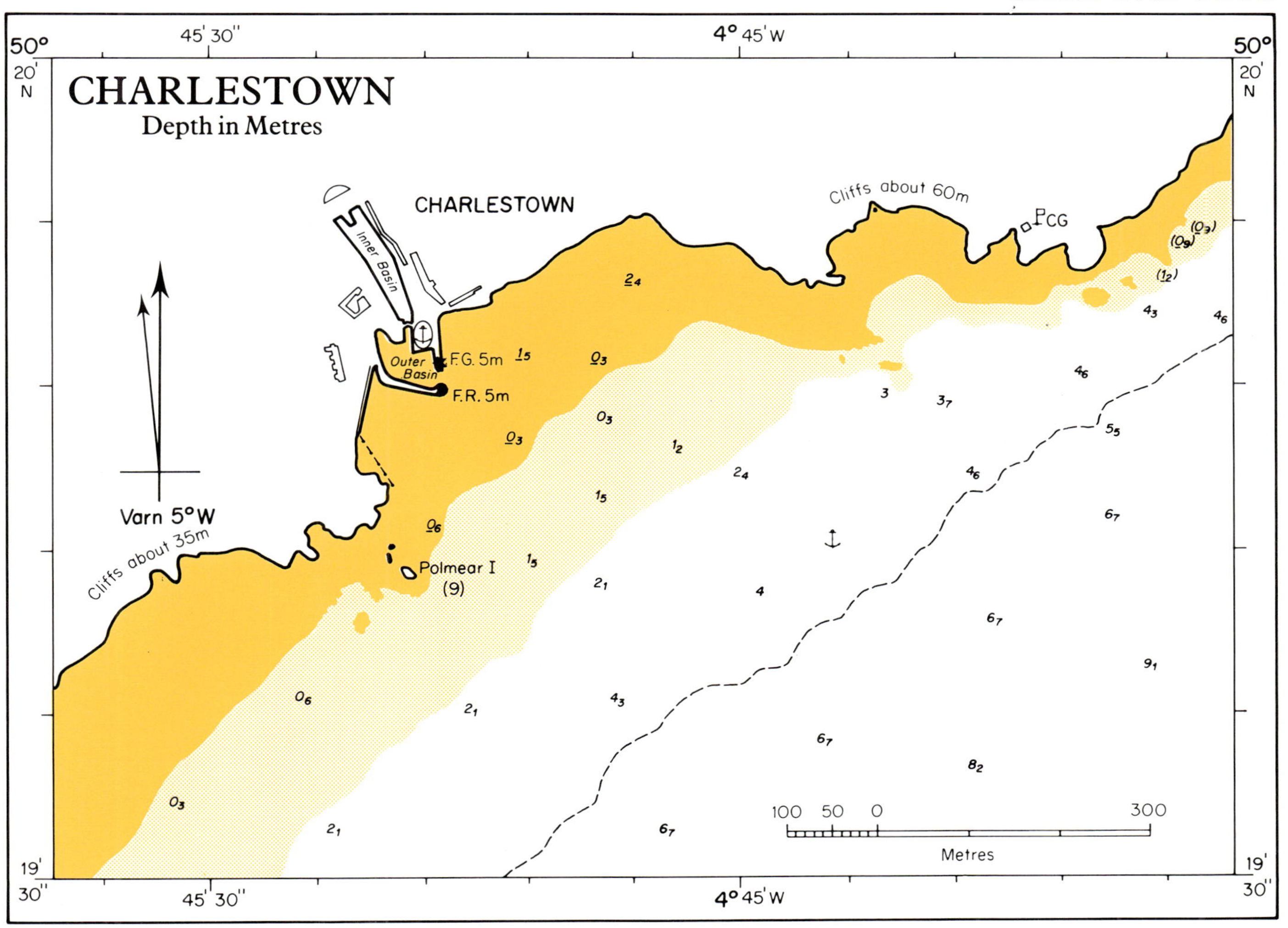

Mevagissey

See Chart No 321.

TIDE

Time and height differences on Standard Port (for instructions in use, see p. xii)

	Lat	Long	Time Difference HW		LW		Height Difference (metres)			
			0000 and 1200	0600 and 1800	0000 and 1200	0600 and 1800				
PLYMOUTH See p. 28	50°22′N	4°11′W					5.5	4.4	2.2	0.8
Mevagissey	50°16′N	4°47′W	−0010	−0015	−0005	+0005	−0.1	−0.1	−0.2	−0.1

TIDES relative to HW Dover (approx) HW −0545. Stream sets E +0200; W−0300.

GENERAL

Mevagissey is a fishing port and popular tourist resort with little space for visiting craft. However it may be possible to lie to two anchors in about 1.5m on the N side of the outer harbour or berth alongside the S pier. The inner harbour dries and is reserved for fishing boats. Good shelter in offshore winds but unsuitable in anything E as a big swell builds up. Boatbuilding at Mevagissey and a yacht yard at Portmellon half a mile S.

APPROACH

The approach is clear and lit.

CAUTIONS

1 Do not approach in offshore winds.
2 Be prepared to move to Fowey or Falmouth if the wind strengthens or shifts E.
3 Speed limit – 3 knots.

TELEPHONES

Harbour Master:	Mevagissey 843305
HM Coastguard:	Mevagissey 842353
MRSC:	Brixham 58292
HM Customs:	Freephone Customs Yachts or (0752) 669811
Medical:	Mevagissey 843701

VHF

Harbour Master: Ch 16; 6; 8; 10; 11; 12; 14; 67; M (portable)

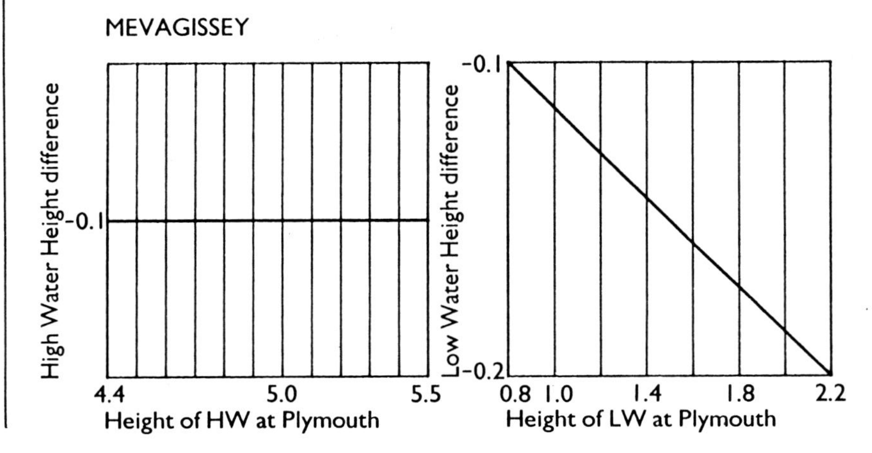

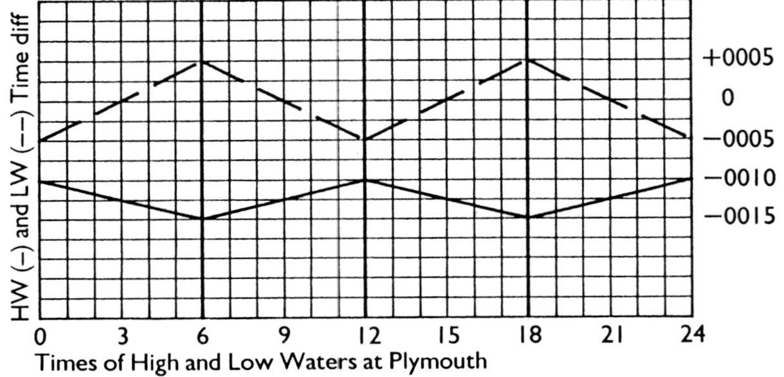

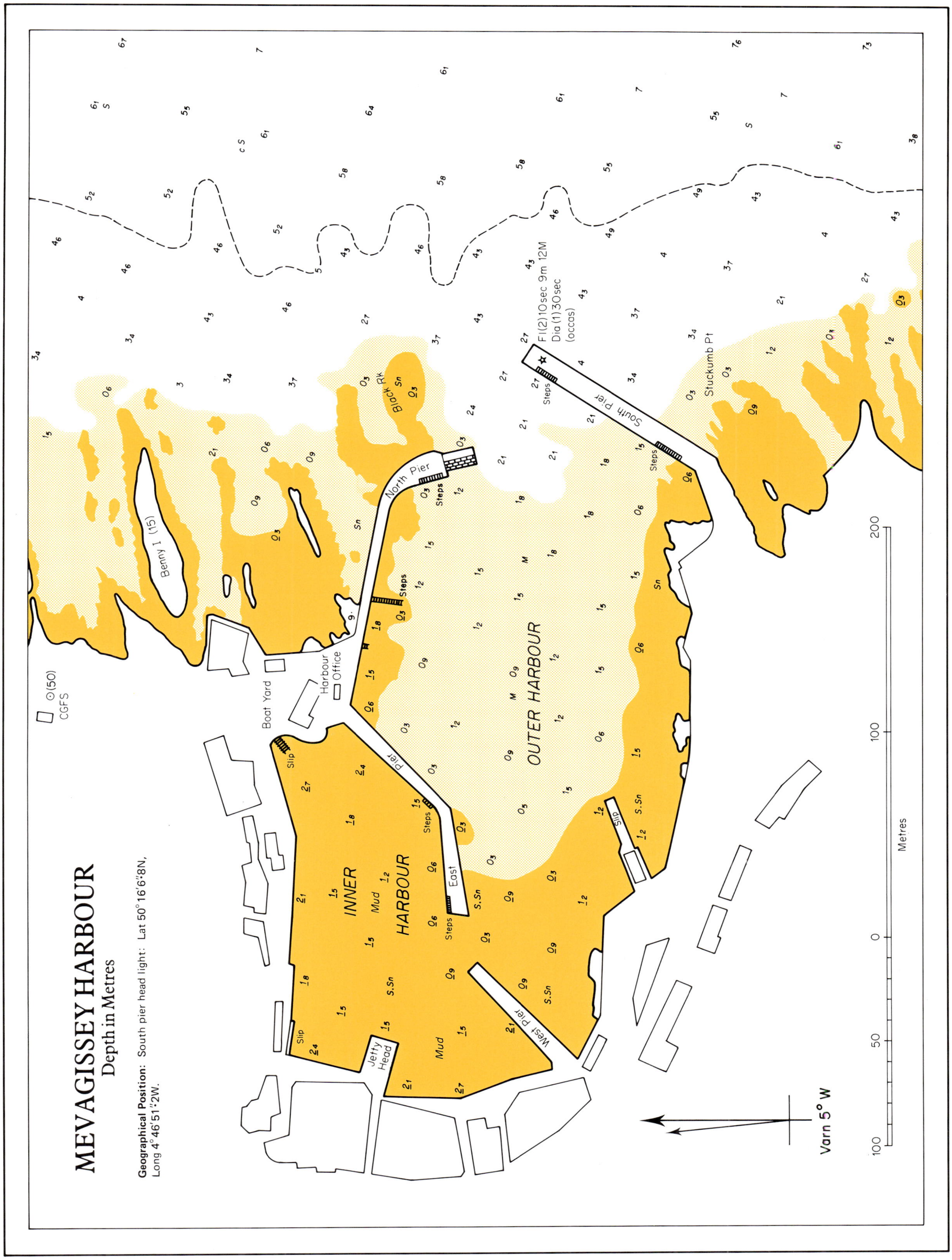
MEVAGISSEY HARBOUR
Depth in Metres
Geographical Position: South pier head light: Lat 50°16′6″8N, Long 4°46′51″2W.
CGFS
⊙(50)
Boat Yard
Harbour Office
Benny I (15)
Black Rk
North Pier
Steps
South Pier
Fl(2)10sec 9m 12M
Dia(1)30sec
(occas)
Stuckumb Pt
OUTER HARBOUR
INNER HARBOUR
Mud
S.Sn
Sn
East
Pier
Slip
Jetty Head
West Pier
Varn 5° W
Metres
100
50
0
100
200

Approaches to Falmouth and St Mawes

See Chart No 322.
See also Falmouth, Mylor and the Truro River on pages 46, 48 and 50.

TIDE

See Falmouth on page 46.

GENERAL

Falmouth harbour and its creeks provide excellent shelter. All facilities.

APPROACH

The approach is easy and entrance may be made in any weather. St Mawes offers all facilities except fuel. There are no visitors' moorings but boats may anchor anywhere clear of moorings and the oyster beds which are laid on both sides of the Percuil River above Polvarth Pt.

CAUTIONS

1 See note above about oyster beds.
2 Large vessels use the port of Falmouth and caution is necessary at the entrance.
3 Black Rock lies in the middle of the entrance but is guarded by an E cardinal lit buoy.
4 Lugo Rock lies on the N side of the entrance to St Mawes but is guarded by a S cardinal unlit buoy.

TELEPHONES

Harbour Master:	St Mawes 750553
Harbour Master:	Falmouth 312285
MRCC:	Falmouth 317575
HM Customs:	Freephone Customs Yachts or (0752) 669811
Medical:	Falmouth 312033 (Doctor) or 315522 (Hospital)

VHF

See Falmouth on page 46.

YACHT CLUB

St Mawes SC:	St Mawes 270686
Percuil SC:	

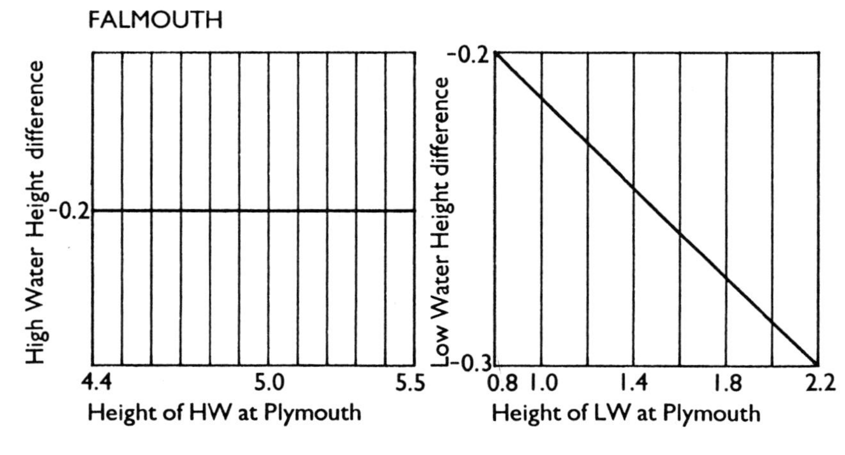

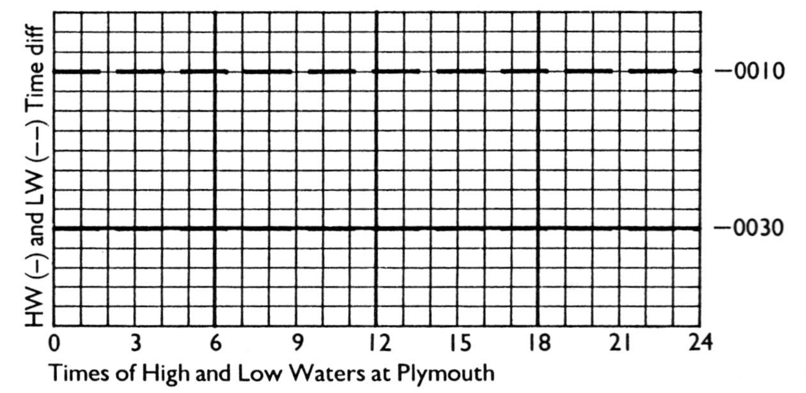

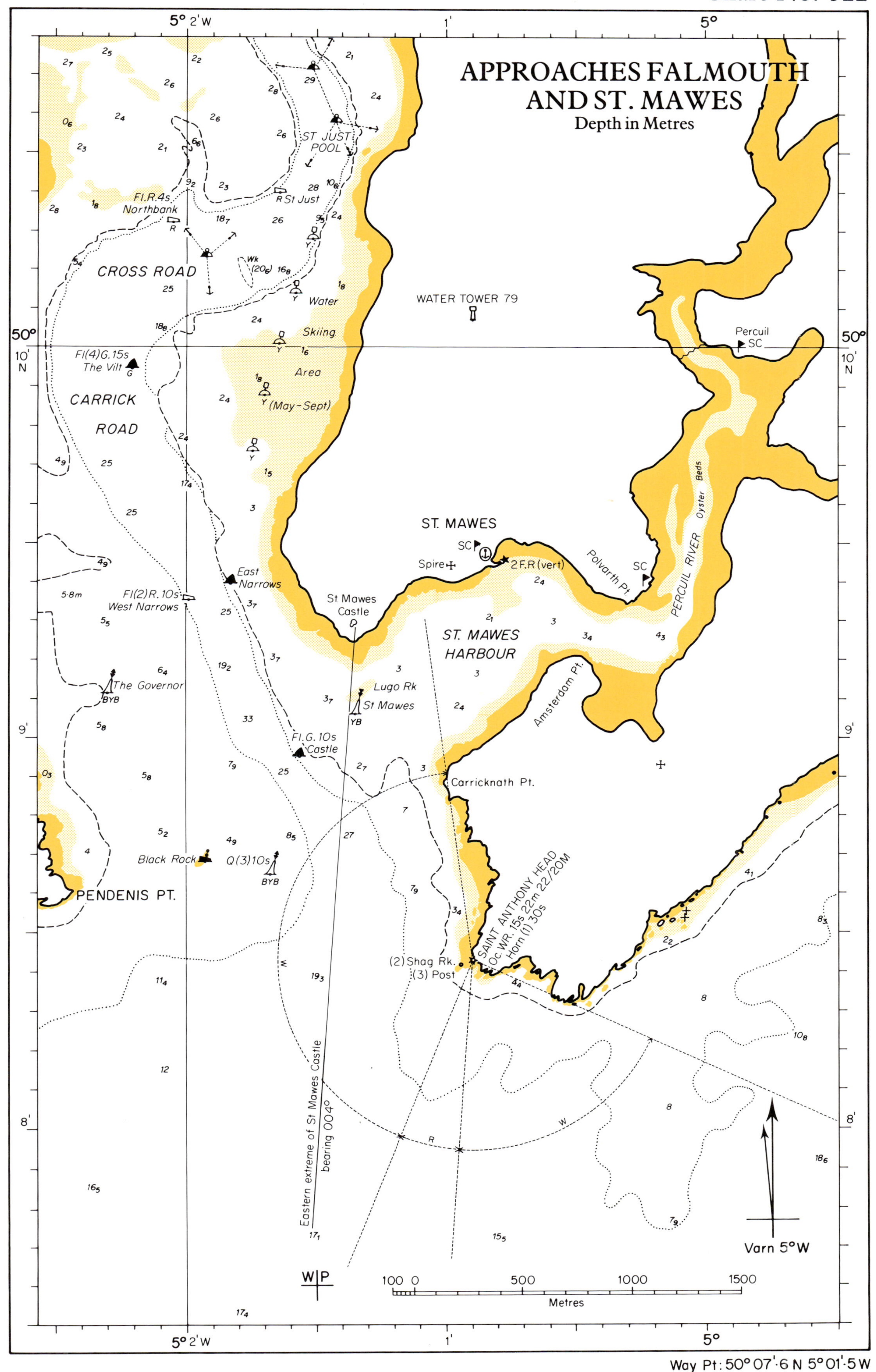
APPROACHES FALMOUTH AND ST. MAWES
Depth in Metres
5° 2'W
1'
5°
50° 10' N
9'
8'
ST JUST POOL
St Just
Fl.R.4s Northbank
CROSS ROAD
CARRICK ROAD
Water Skiing Area
Y (May-Sept)
Wk (20₆)
WATER TOWER 79
Percuil SC
Fl(4)G.15s The Vilt
East Narrows
Fl(2)R.10s West Narrows
The Governor
BYB
ST. MAWES
SC
Spire
2F.R(vert)
Polvarth Pt
PERCUIL RIVER
Oyster Beds
St Mawes Castle
ST. MAWES HARBOUR
Amsterdam Pt.
Lugo Rk
St Mawes
YB
Fl.G.10s Castle
Carricknath Pt.
Black Rock
Q(3)10s
PENDENIS PT.
SAINT ANTHONY HEAD Oc.WR.15s 22m 22/20M Horn (1) 30s
(2) Shag Rk.
(3) Post
Eastern extreme of St Mawes Castle bearing 004°
W|P
100 0 500 1000 1500
Metres
Varn 5°W

Way Pt: 50° 07'·6 N 5° 01'·5 W

Falmouth

See Chart No 323.

See also Approaches to Falmouth, Mylor and the Truro River on pages 44, 48 and 50.

TIDE

Time and height differences on Standard Port (for instructions in use, see p. xii)

	Lat	Long	Time Difference HW		Time Difference LW		Height Difference (metres)			
PLYMOUTH See p. 28	50°22′N	4°11′W	0000 and 1200	0600 and 1800	0000 and 1200	0600 and 1800	5.5	4.4	2.2	0.8
Falmouth	50°09′N	5°03′W	−0030	−0030	−0010	−0010	−0.2	−0.2	−0.3	−0.2

TIDES relative to HW Dover (approx) HW +0605. Stream sets W −0300; E +0200.

GENERAL

Falmouth offers all facilities for yachts, two marinas and a number of visitors' moorings. There are large shipbuilding and repairing yards.

APPROACH

See Approaches to Falmouth on page 44.

CAUTIONS

1 Very large vessels may be using the docks.
2 The bottom of the harbour off the town is foul; buoy the anchor.

TELEPHONES

Harbour Master:	Falmouth 312285
MRCC:	Falmouth 317575
HM Customs:	Freephone Customs Yachts or (0752) 669811
Medical:	Falmouth 312033 (Doctor) or 315522 (Hospital)

VHF

Falmouth Hr Radio:	Ch 16; 12; 13; 14
Customs Launch:	Ch 16; 6; 9; 10; 12; 14
Falmouth Yacht Marina:	Ch M
R Cornwall YC:	Ch M

MARINAS

Falmouth Yacht Marina:	Falmouth 316620
Visitors' Yacht Haven:	Falmouth 312285

YACHT CLUBS

R Cornwall YC:	Falmouth 312126
Flushing SC:	Falmouth 74043

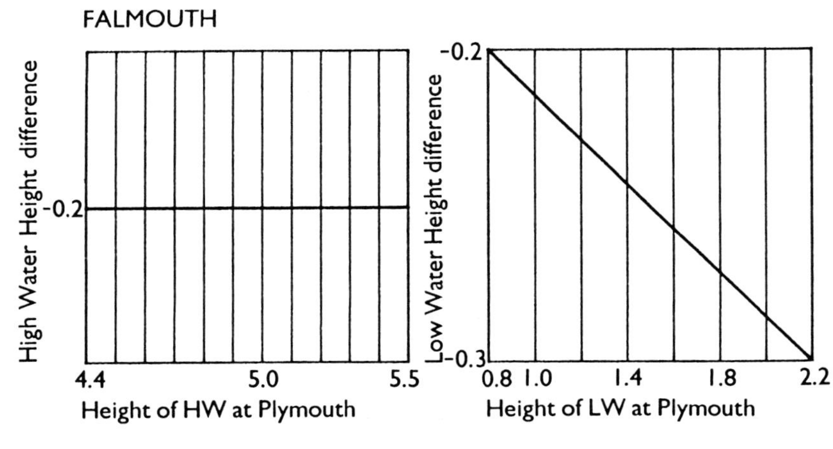

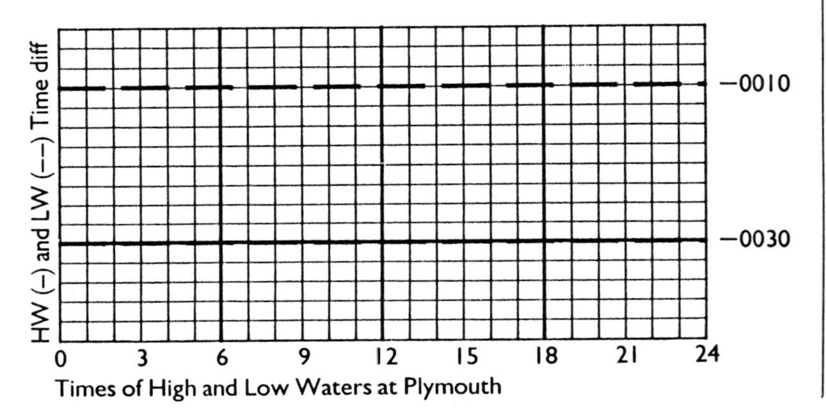

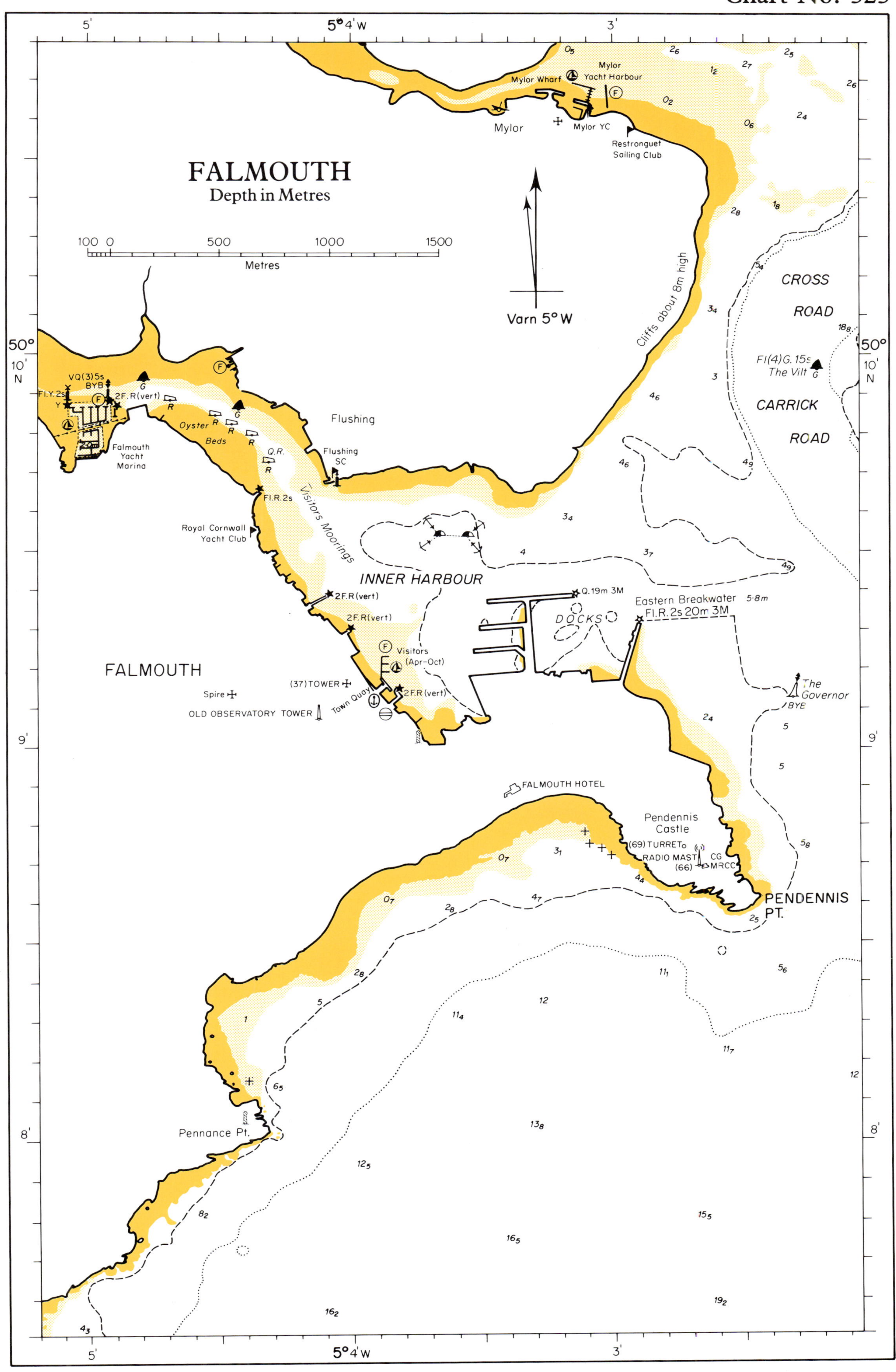
FALMOUTH
Depth in Metres
Metres
Varn 5° W
Mylor
Mylor Wharf
Mylor Yacht Harbour
Mylor YC
Restronguet Sailing Club
Cliffs about 8m high
CROSS
ROAD
CARRICK
ROAD
Fl(4)G.15s The Vilt
Flushing
Flushing SC
Oyster Beds
Visitors Moorings
Falmouth Yacht Marina
Royal Cornwall Yacht Club
INNER HARBOUR
DOCKS
Eastern Breakwater
Fl.R.2s 20m 3M
Q.19m 3M
The Governor BYE
FALMOUTH
Visitors (Apr-Oct)
(37) TOWER
Spire
OLD OBSERVATORY TOWER
Town Quay
FALMOUTH HOTEL
Pendennis Castle
(69) TURRET
RADIO MAST (66)
CG MRCC
PENDENNIS PT.
Pennance Pt.

Mylor and Restronguet

See Chart No 324.

See also Approaches to Falmouth, Falmouth and Truro River on pages 44, 46 and 50.

TIDE

See Falmouth on page 46.

GENERAL

This middle part of Falmouth Harbour offers excellent shelter in the numerous creeks and bays. All facilities at Mylor.

APPROACH

The approach is easy and entrance may be made in any weather. St Mawes offers all facilities except fuel. There are no visitors' moorings but boats may anchor anywhere clear of moorings and the oyster beds which are laid on both sides of the Percuil River above Polvarth Pt.

TELEPHONES

Harbour Master:	St Mawes 750553
Harbour Master:	Falmouth 312285
MRCC:	Falmouth 317575
HM Customs:	Freephone Customs Yachts or (0752) 669811
Medical:	Falmouth 312033 (Doctor) or 315522 (Hospital)

VHF

Mylor Yacht Harbour:	Ch M

MARINA

Mylor Yacht Harbour:	Falmouth 72121

YACHT CLUBS

Mylor YC:	Falmouth 74391
Restronguet SC:	

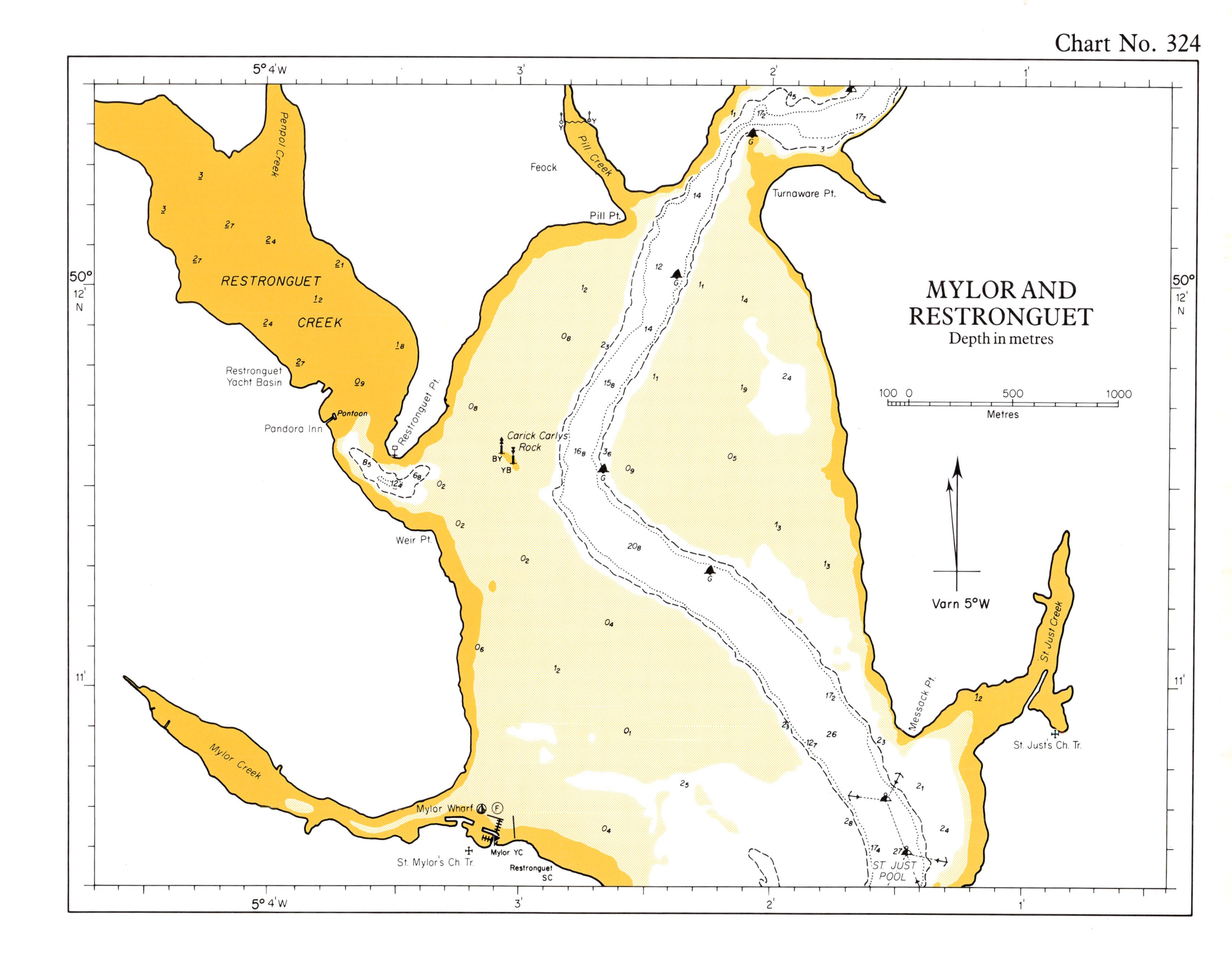
MYLOR AND RESTRONGUET
Depth in metres
100 0 500 1000
Metres
Varn 5°W
5° 4'W
3'
2'
1'
50° 12' N
11'
Penpol Creek
RESTRONGUET CREEK
Restronguet Yacht Basin
Pontoon
Pandora Inn
Restronguet Pt.
Feock
Pill Creek
Pill Pt.
Turnaware Pt.
Carick Carlys Rock
BY
YB
Weir Pt.
Mylor Creek
Mylor Wharf
St. Mylor's Ch. Tr.
Mylor YC
Restronguet SC
Messack Pt.
St. Just Creek
St. Just's Ch. Tr.
ST. JUST POOL

Truro River

See Chart No 325.

See also Approaches to Falmouth and St Mawes, Falmouth and Mylor on pages 44, 46 and 48.

TIDE

Time and height differences on Standard Port (for instructions in use, see p. xii)

	Lat	Long	Time Difference HW		LW		Height Difference (metres)			
			0000 and 1200	0600 and 1800	0000 and 1200	0600 and 1800				
PLYMOUTH See p. 28	50°22′N	4°11′W					5.5	4.4	2.2	0.8
Truro	50°16′N	5°03′W	−0020	−0025	dries		−2.0	−2.0	dries	

GENERAL

The upper reaches of Falmouth Harbour lead to the Truro River which is navigable on tide as far as the city of Truro. There are often large merchant vessels laid up in King Harry Passage. There is a marina about halfway up the river.

APPROACH

Truro Marina is accessible −4HW+4. The river shoals rapidly after passing the laying up moorings.

CAUTIONS

1 There are concrete mooring blocks which dry 3 cables S and half a cable N of the King Harry ferry crossing.
2 There may be other obstructions in the entrance to Ruan Creek.

TELEPHONES

Harbour Master:	Truro 78131
Harbour Master:	Falmouth 312285
MRCC:	Falmouth 317575
HM Customs:	Freephone Customs Yachts or (0752) 669811
Medical:	Falmouth 312033 (Doctor) or 315522 (Hospital)

VHF

Falmouth Hr Radio:	Ch 16; 12; 13; 14
Customs Launch:	Ch 16; 6; 9; 10; 12; 14
Falmouth Yacht Marina:	Ch M
R Cornwall YC:	Ch M

MARINA

Truro Marina: Truro 79854

TRURO RIVER

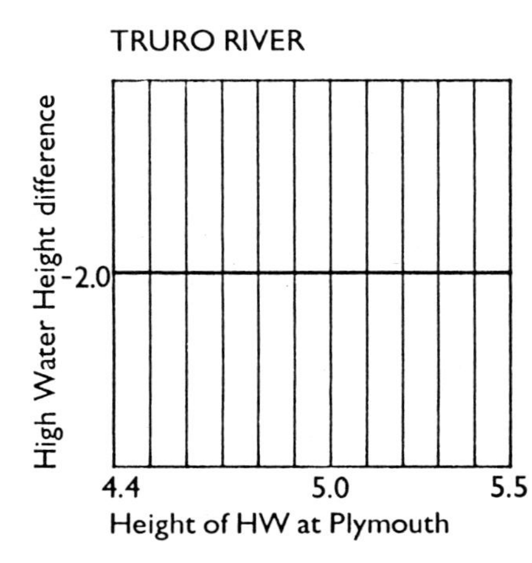

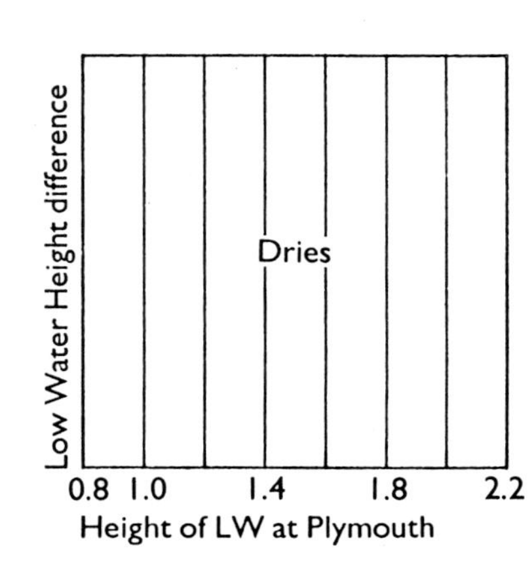

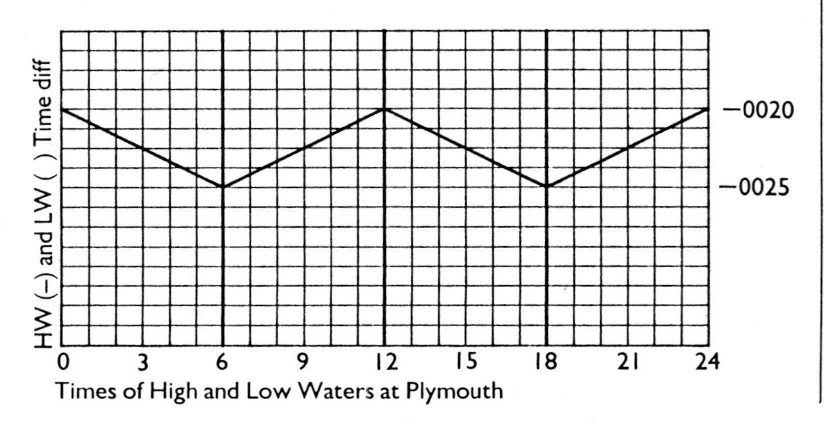

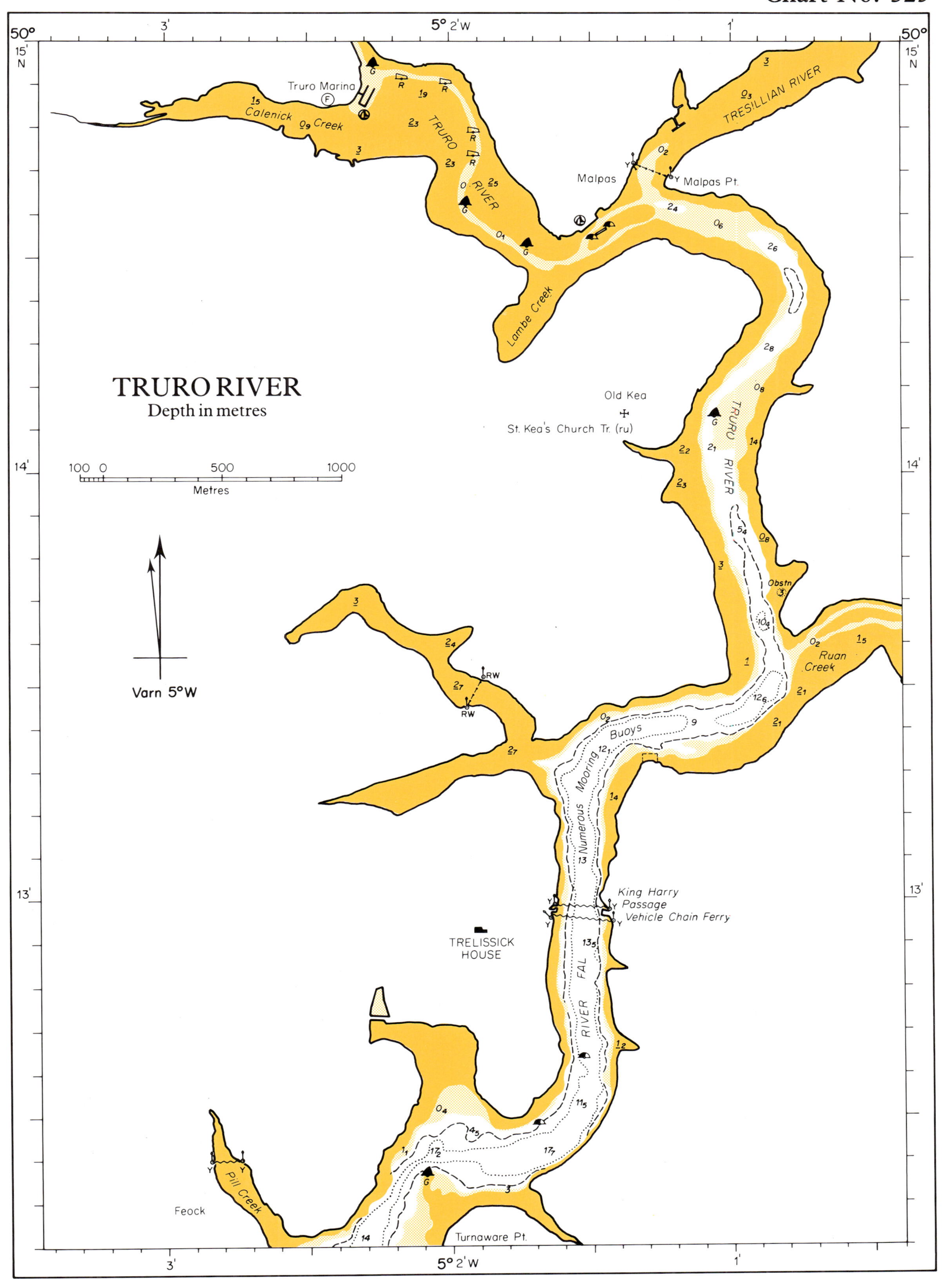
TRURO RIVER
Depth in metres
100 0 500 1000
Metres
Varn 5°W
Truro Marina
Calenick Creek
TRURO RIVER
TRESILLIAN RIVER
Malpas
Malpas Pt.
Lambe Creek
Old Kea
St. Kea's Church Tr. (ru)
TRURO RIVER
Obstn
Ruan Creek
Numerous Mooring Buoys
King Harry Passage
Vehicle Chain Ferry
TRELISSICK HOUSE
RIVER FAL
Pill Creek
Feock
Turnaware Pt.
5° 2'W
3'
1'
50° 15' N
14'
13'

Helford River

See Chart No 326.

TIDE

Time and height differences on Standard Port (for instructions in use, see p. xii)

	Lat	Long	Time Difference HW		LW		Height Difference (metres)			
			0000 and 1200	0600 and 1800	0000 and 1200	0600 and 1800				
PLYMOUTH See p. 28	50°22′N	4°11′W					5.5	4.4	2.2	0.8
Helford Entr	50°05′N	5°05′W	−0030	−0035	−0015	−0010	−0.2	−0.2	−0.3	−0.2

TIDES relative to HW Dover (approx) HW −0610. Stream turns 0100 after local HW and LW.

GENERAL

Helford provides a beautiful, quiet and sheltered harbour, except in strong E winds, with plenty of water for most yachts and an easy, though unlit, entrance. All facilities.

APPROACH

The only danger is the Gedges Rock on the N side of the entrance. Anchor as convenient, keeping clear of oyster beds in Navas Creek and W of its entrance. There are also visitors' buoys off Helford and visitors' pontoons in Navas Creek.

CAUTIONS

1 Keep clear of oyster beds.
2 Though there is generally plenty of water, bars exist at the entrance to Navas Creek and the mud bank off Helford is very steep.
3 Speed limit – 6 knots.

TELEPHONES

Harbour Master:	St Keverne 280422
HM Coastguard:	St Keverne 280221
MRCC:	Falmouth 317575
HM Customs:	Freephone Customs Yachts or (0752) 669811
Medical:	Helston 2151 (Hospital)

VHF

Helford River SC:	Ch M
Gweek Quay (yard and supplies):	Ch M

YACHT CLUB

Helford River SC:	Manaccan 460

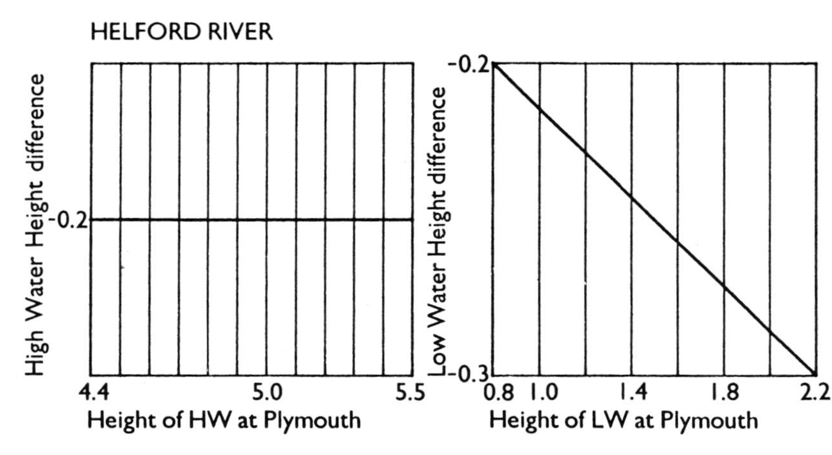

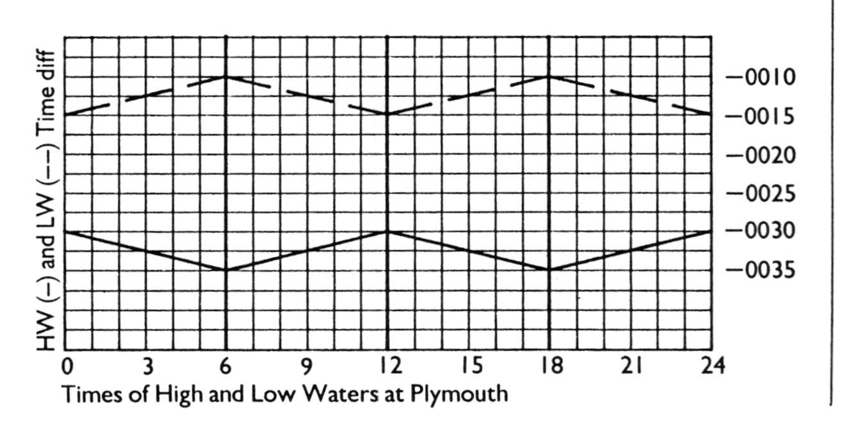

Chart No. 326

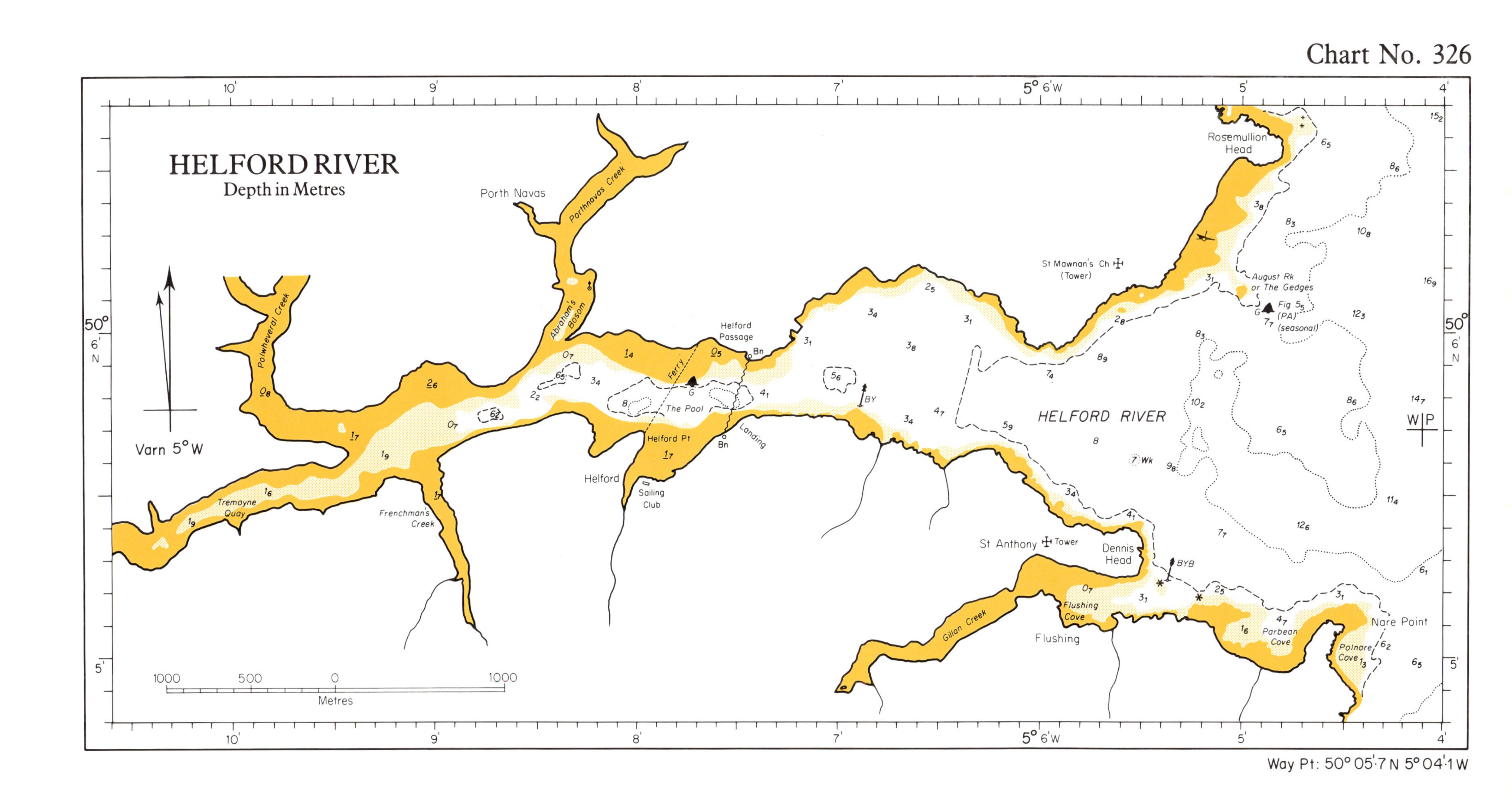

HELFORD RIVER
Depth in Metres
Varn 5° W
Porth Navas
Porthnavas Creek
Polwheveral Creek
Abraham's Bosom
Tremayne Quay
Frenchman's Creek
Helford
Sailing Club
Helford Pt
Helford Passage
Ferry
The Pool
Landing
Bn
BY
St Mawnan's Ch (Tower)
Rosemullion Head
August Rk or The Gedges
Fig 5s (PA) (seasonal)
HELFORD RIVER
Wk
WP
St Anthony Tower
Dennis Head
BYB
Flushing Cove
Flushing
Gillan Creek
Parbean Cove
Polnare Cove
Nare Point
Metres
1000 500 0 1000
Way Pt: 50° 05'·7 N 5° 04'·1 W

Penzance

See Chart No 327.

TIDE

Time and height differences on Standard Port (for instructions in use, see p. xii)

	Lat	Long	Time Difference HW		Time Difference LW		Height Difference (metres)			
PLYMOUTH See p. 28	50°22′N	4°11′W	0000 and 1200	0600 and 1800	0000 and 1200	0600 and 1800	5.5	4.4	2.2	0.8
Penzance	50°06′N	5°33′W	−0040	−0105	−0045	−0020	+0.1	0.0	−0.2	0.0

TIDES relative to HW Dover (approx) HW +0600.

GENERAL

Penzance offers good shelter within a wet dock. There is also a drying harbour. The ferry to the Scilly Isles uses the Albert Pier but she berths overnight on the Lighthouse Pier. Most facilities for yachts.

APPROACH

The W sector is clear of dangers but the W extremity comes close to the Gear Rock (dries 1.8m, see Chart 328). Anchor off to the NNE of the entrance or make fast alongside either the Albert or Lighthouse Piers if space is available until the dock gates open (−2HW+1). Traffic signals on FS on N arm of dock: 2 balls (hor) or 2 R Lts (vert) – dock open; 2 balls (vert) or R Lt over G Lt – dock closed. Wait until signalled to proceed by the dockmaster who will allocate a berth.

There is an Aero Radio DF Beacon as shown, 'PH', 333 kHz, 15 M (Mon–Sat 0700–1930).

CAUTIONS

1. Mount's Bay is open to SE winds and Penzance should not be approached in strong winds from this quarter.
2. The entrance shoals on the N side.
3. In S gales, seas break over the S arm.
4. There is a speed limit of 5 knots within the harbour.
5. There is a ledge about 1m wide all round the inside of the wet dock with 1.4m at MLWS.

TELEPHONES

Harbour Master:	Penzance 61119
HM Coastguard:	Penzance 87351
MRCC:	Falmouth 31575
HM Customs:	Freephone Customs Yachts or (0752) 669811
Medical:	Penzance 63866 (Doctor)

VHF

Penzance Harbour:	Ch 16; 9; 12 (−2HW+1 and office hours)

YACHT CLUB

Penzance SC:	Penzance 64989

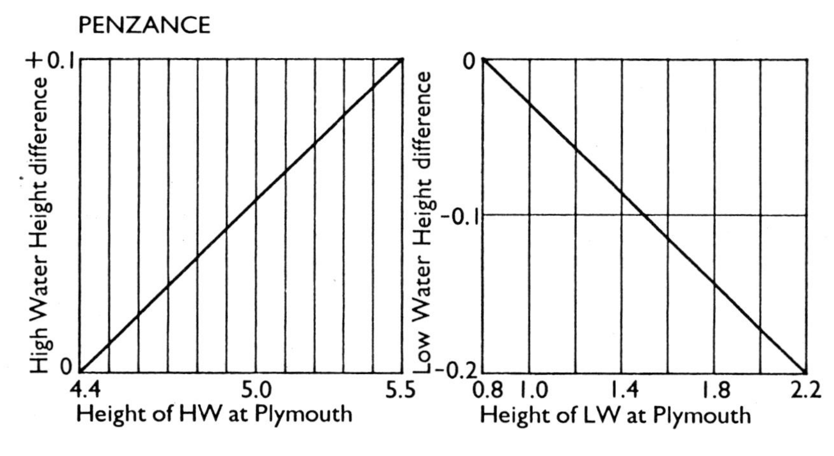

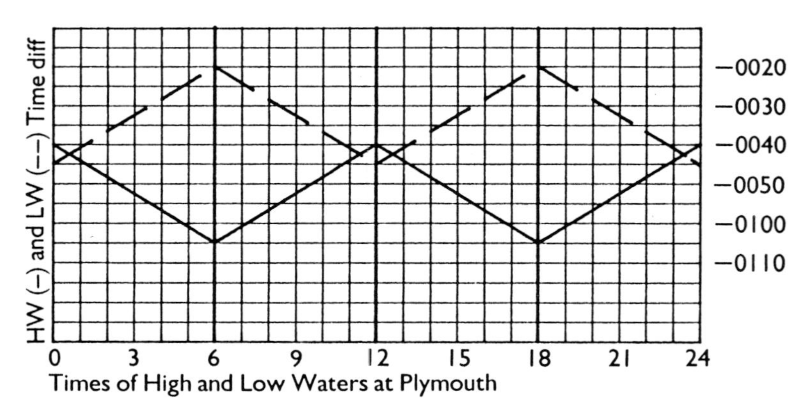

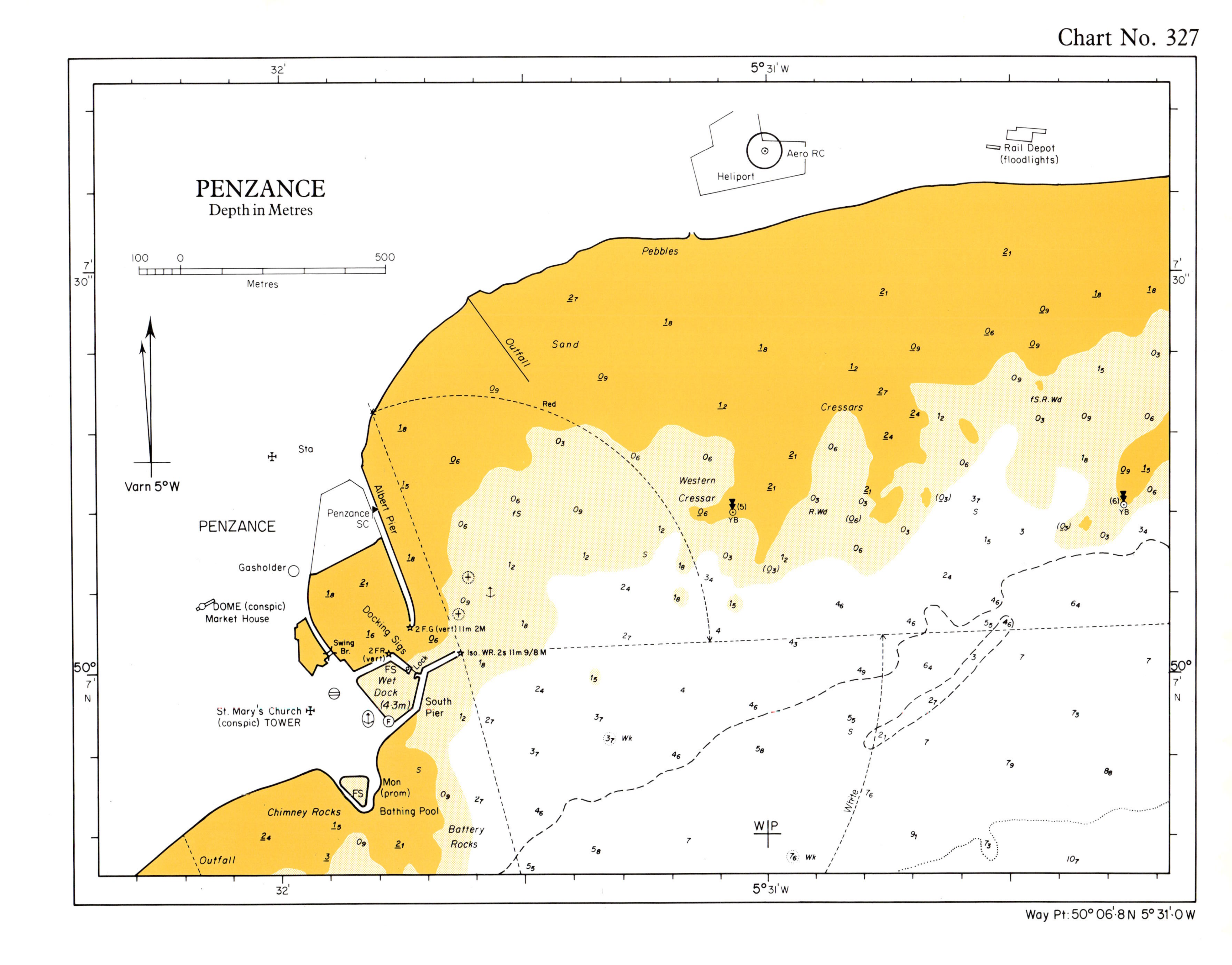

Way Pt: 50° 06'·8 N 5° 31'·0 W

Newlyn

See Chart No 328.

TIDE

Time and height differences on Standard Port (for instructions in use, see p. xii)

	Lat	Long	Time Difference HW		LW		Height Difference (metres)			
PLYMOUTH See p. 28	50°22′N	4°11′W	0000 and 1200	0600 and 1800	0000 and 1200	0600 and 1800	5.5	4.4	2.2	0.8
Newlyn	50°06′N	5°33′W	−0040	−0105	−0045	−0020	+0.1	0.0	−0.2	0.0

TIDES relative to HW Dover (approx) HW +0600.

GENERAL

Newlyn is a fishing port with some commercial activity. It provides good shelter but should only be used for an overnight stay except by arrangement with the Harbour Master. Most facilities. There is over 2m depth at the entrance and head of the N and S piers. The quay which extends parallel to the N pier is dredged to 1.8m on either side.

APPROACH

The approach is clear of dangers except for the Gear Rock to the NE. The S pier is used by coasters and the N pier by fishing boats. Unless told otherwise, berth outside a fishing boat but be prepared to move at short notice.

CAUTION

1 There is a speed limit of 3 knots.

TELEPHONES

Harbour Master:	Penzance 62523
HM Coastguard:	Penzance 87351
MRCC:	Falmouth 317575
HM Customs:	Freephone Customs Yachts or (0752) 669811
Medical:	Penzance 63866 (Doctor)

VHF

Newlyn Harbour:	Ch 16; 12 (office hrs)

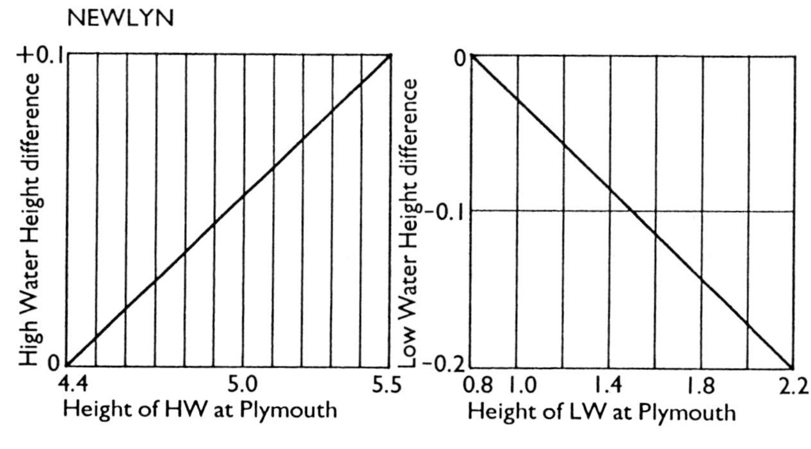

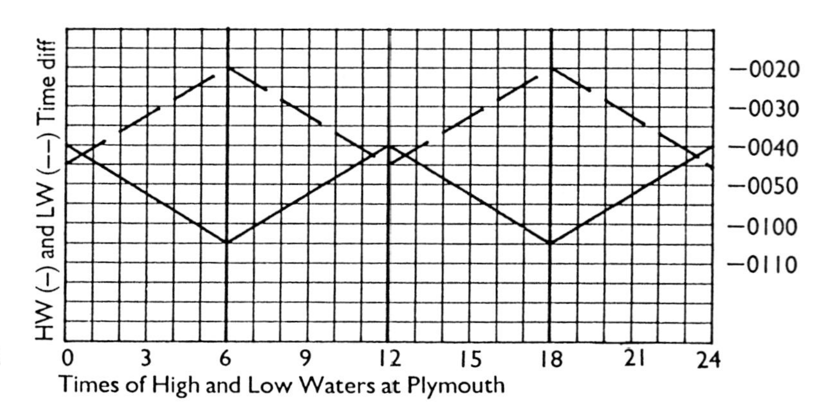

Chart No. 328

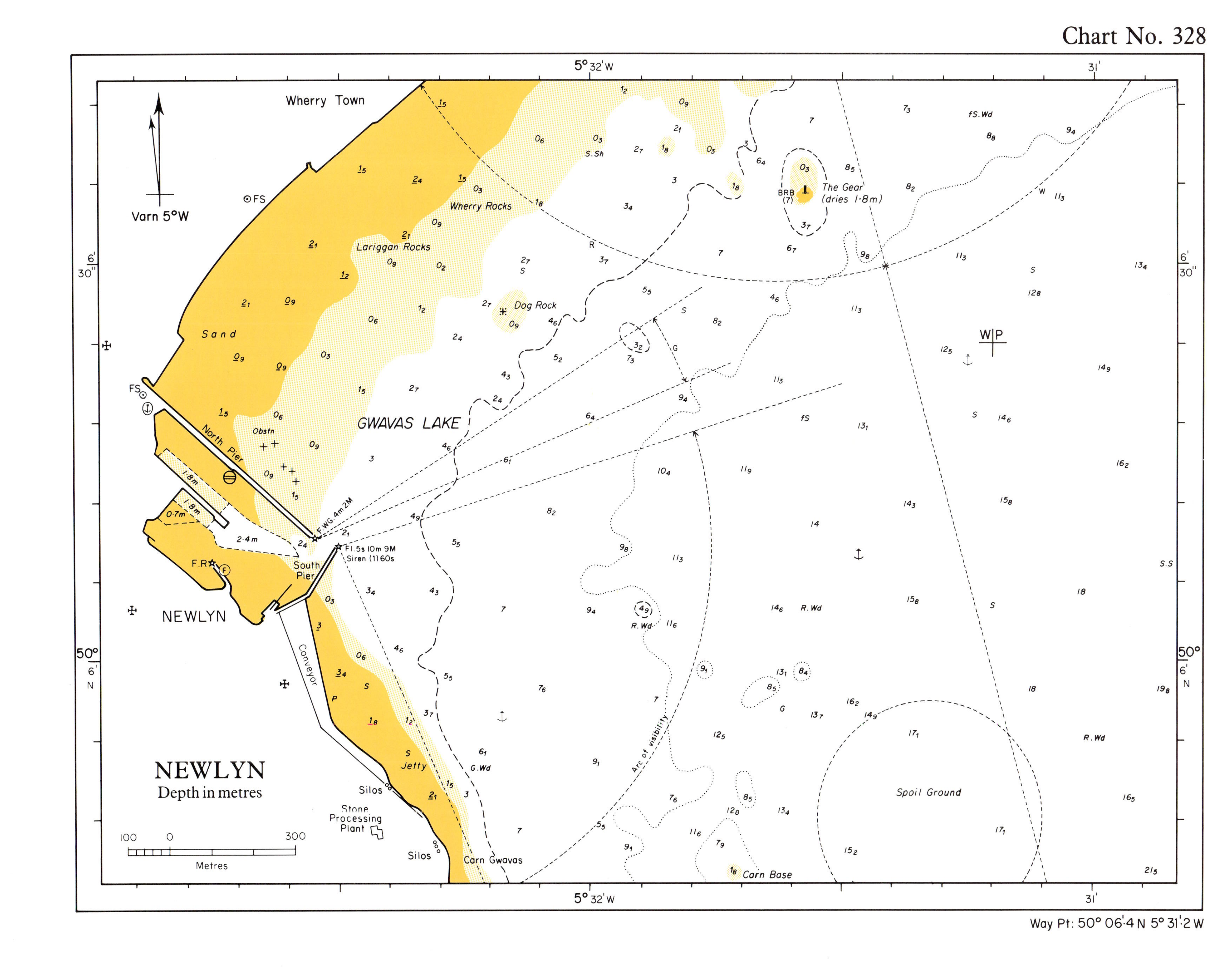

NEWLYN
Depth in metres
Wherry Town
Varn 5°W
Wherry Rocks
Lariggan Rocks
Sand
Dog Rock
GWAVAS LAKE
The Gear (dries 1·8m)
BRB (7)
North Pier
South Pier
Obstn
F.WG.4m 2M
Fl.5s 10m 9M Siren (1) 60s
F.R
NEWLYN
Conveyor
Jetty
Silos
Stone Processing Plant
Carn Gwavas
Carn Base
Spoil Ground
Arc of visibility
W|P
5° 32'W
31'
6' 30"
50° 6' N
Metres
100 0 300
Way Pt: 50° 06'4 N 5° 31'2 W

Mousehole

See Chart No 329.

TIDE

See page 56.

GENERAL

Mousehole is a tiny, drying harbour with a narrow (11m wide) entrance. It may be entered in good weather for curiosity value but will probably be too crowded to berth. Tourist facilities.

APPROACH

Approach from the S. There is a good holding between St Clement's Isle and the harbour on the 5m sounding. Passage to the N is encumbered by unmarked rocks.

CAUTION

1 In strong onshore winds the harbour entrance is closed in which event a R lt replaces the 2 FG (vert) on the N pier head.

TELEPHONES

Harbour Master:	Penzance 731511
HM Coastguard:	Penzance 87351
MRCC:	Falmouth 317575

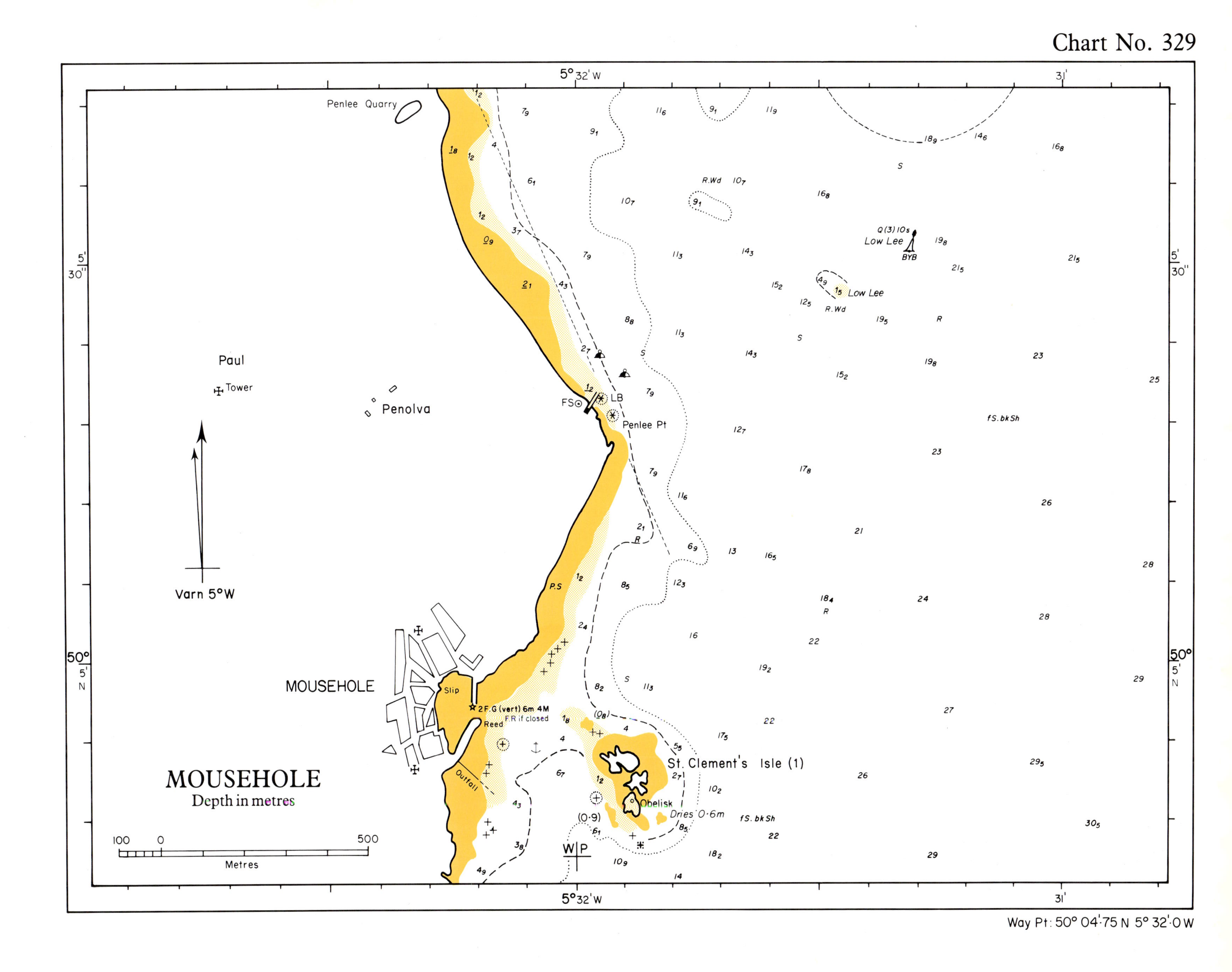

Way Pt: 50° 04'·75 N 5° 32'·0 W

St Mary's and St Agnes (Isles of Scilly) See Chart No 330.

TIDE

Time and height differences on Standard Port (for instructions in use, see p. xii)

	Lat	Long	Time Difference HW		LW		Height Difference (metres)			
PLYMOUTH See p. 28	50°22′N	4°11′W	0000 and 1200	0600 and 1800	0000 and 1200	0600 and 1800	5.5	4.4	2.2	0.8
St Mary's	49°55′N	6°19′W	−0030	−0110	−0100	−0020	+0.2	−0.1	−0.2	−0.1

TIDES relative to HW Dover (approx) HW +0600.

GENERAL

The Isles of Scilly are beautiful and provide a wonderful cruising ground in settled weather. Most facilities at St Mary's.

APPROACH

Great care is needed to approach and to sail within the Isles. Apart from the principal lights there are few others and pilotage is achieved by leading lines which require daylight. The principal anchorages are marked. There is an Aero Radio DF Beacon, 'STM', as shown, 321kHz, 15M.

CAUTIONS

1 There are many rock pinnacles.
2 The tides run strongly with confusing eddies and some overfalls.

TELEPHONES

(St Mary's) Harbour Master:	Scillonia 22768
HM Coastguard:	Scillonia 22651
MRCC:	Falmouth 317575
HM Customs:	Scillonia 22571
Medical:	Scillonia 22628 (Doctor)

VHF

St Mary's Harbour: Ch 16; 14 (office hours)

YACHT CLUB

Isles of Scilly YC: Scillonia 22069

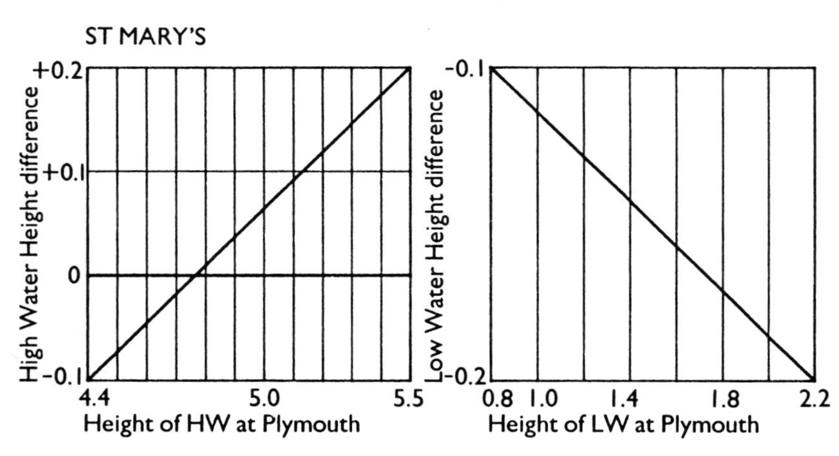

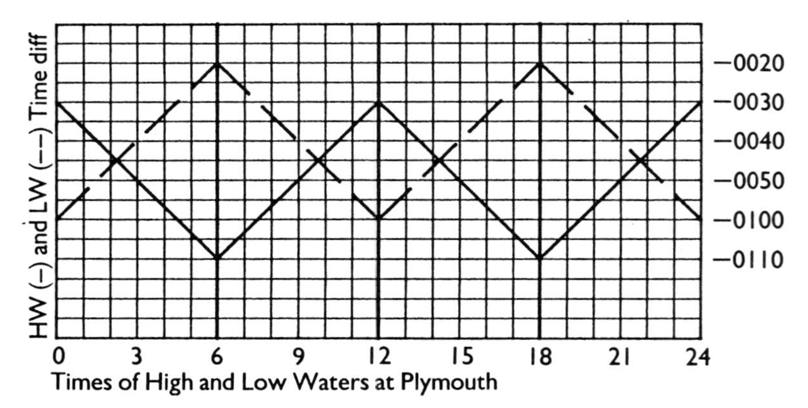

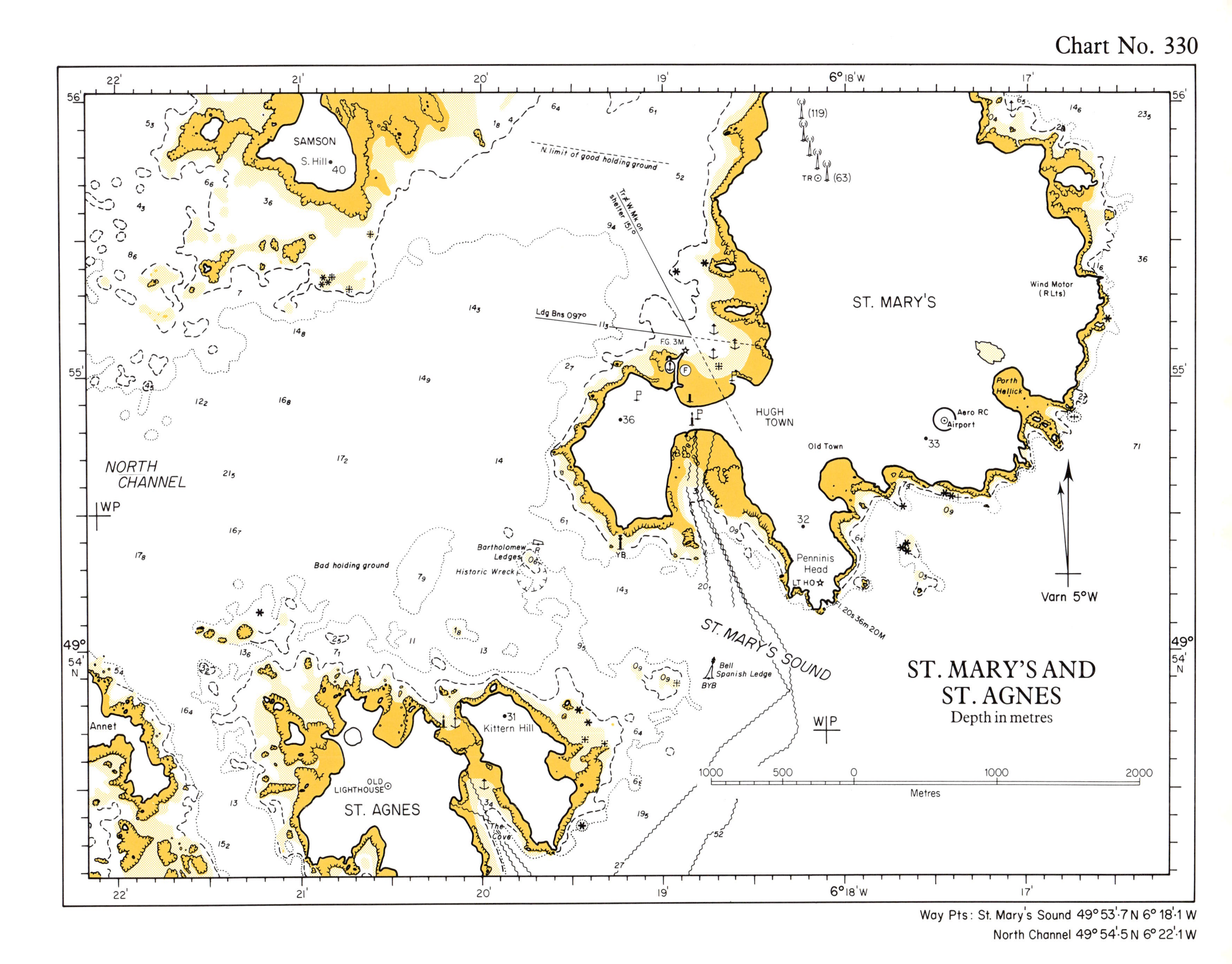
ST. MARY'S AND
ST. AGNES
Depth in metres
ST. MARY'S
HUGH TOWN
Old Town
Porth Hellick
Aero RC
Airport
Wind Motor (R Lts)
Penninis Head
LT HO
20s 36m 20M
FG. 3M
Ldg Bns 097°
Tr ≠ W. Mk on shelter 151°
N. limit of good holding ground
SAMSON
S. Hill 40
NORTH CHANNEL
ST. MARY'S SOUND
Bell Spanish Ledge BYB
Bartholomew Ledges
Historic Wreck
Bad holding ground
ST. AGNES
OLD LIGHTHOUSE
Kittern Hill
The Cove
Annet
Varn 5°W
WP
Metres
1000 500 0 1000 2000
6°18'W
49° 54' N
Way Pts: St. Mary's Sound 49°53'·7 N 6°18'·1 W
North Channel 49°54'·5 N 6°22'·1 W

Tresco (Isles of Scilly)

See Chart No 331.

TIDE

See page 60.

GENERAL

The Isles of Scilly are beautiful and provide a wonderful cruising ground in settled weather. Though Tresco has famous gardens, most facilities for yachtsmen are to be found at St Mary's.

APPROACH

Great care is needed to approach and to sail within the Isles. Apart from the principal lights there are few others and pilotage is achieved by leading lines which require daylight. The principal anchorages are marked. There is a Radio DF Beacon on Round Island, 'RR', 308kHz, 100M.

CAUTIONS

1 There are many rock pinnacles.
2 The tides run strongly with confusing eddies and some overfalls.

TELEPHONES

(St Mary's)

Harbour Master:	Scillonia 22768
HM Coastguard:	Scillonia 22651
MRCC:	Falmouth 317575
HM Customs:	Scillonia 22571
Medical:	Scillonia 22628 (Doctor)

TRESCO

Depth in metres

Varn 5°W

LIGHTHOUSE
Fl. 10s 55m 24M
Siren (4) 60s

Round I.
RC

1000 500 0 1000
Metres

W|P

Men-a-vaur

OLD GRIMSBY SOUND

St. Helen's
42
Landing Carn

St Helen's Pool

Tean

Tean Sound

White I.

St. Martin's Head

(56) RW DAYMARK

ST. MARTIN'S

Beacon Hill (42) Ru

Old Grimsby

BRYHER

DAYMARK

Quay

Slip

New Grimsby

TRESCO

Abbey Hill Mon (38)

Tresco Abbey

Great Rock

TV Tower

Bn 160° 30'

Centre of Men-a-vaur in line with Landing Carn 322°

TRESCO FLATS

CROW BAR

BRB

(0₆) Boiler

Hats

YB

CROW SOUND

Great Gahilly

Great Arthur

W|P

6°21'W 20' 19' 18' 17' 16'

58' 49° 57' N 56'

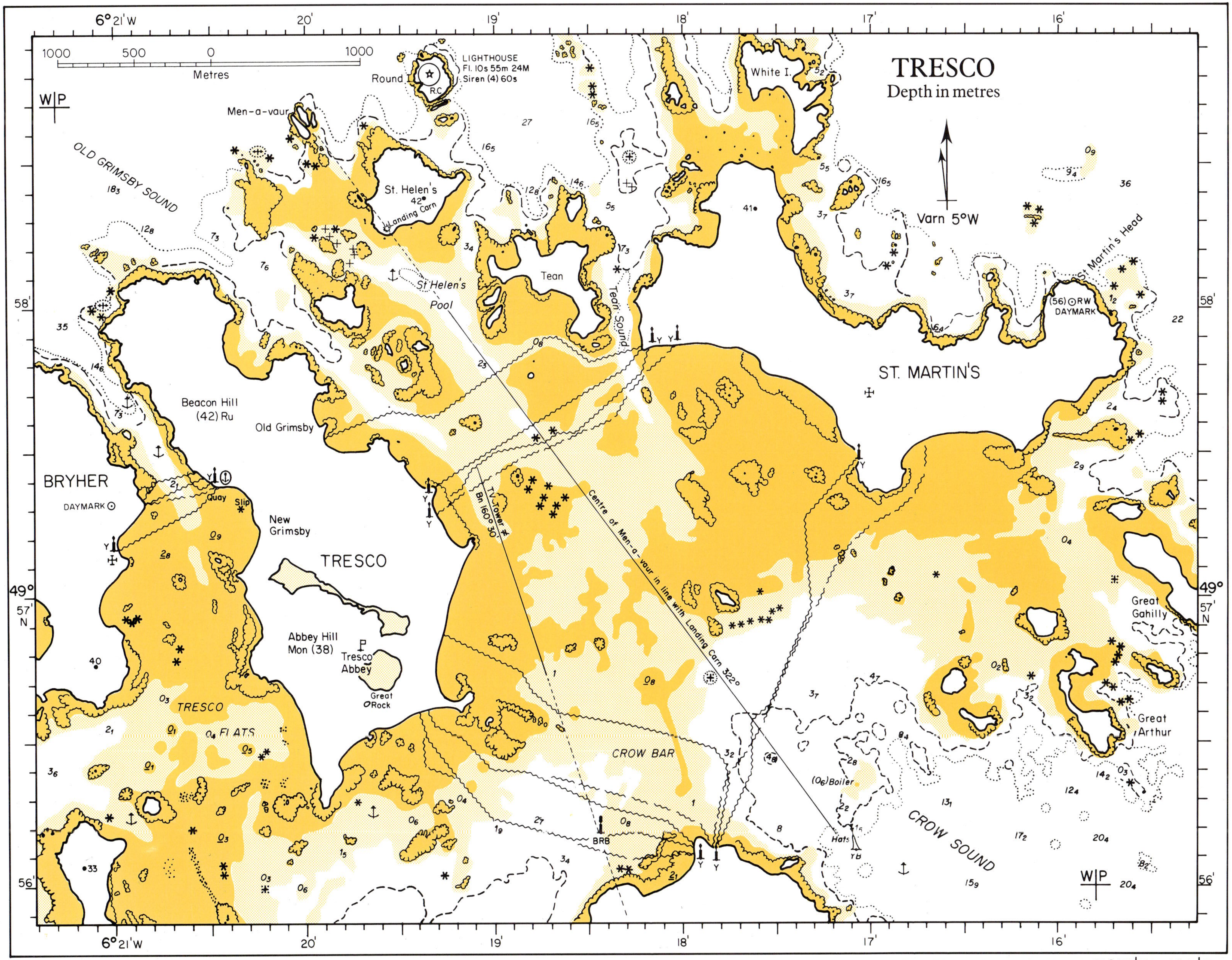

Way Pts: Old Grimsby Sound 49° 58'·7 N 6° 21'·3 W
Crow Sound 49° 56'·0 N 6° 15'·8 W

Le Légué and St Brieuc

See Chart No 332.

TIDE

Time and height differences on Standard Port (for instructions in use, see p. xii)
Note that this table, using times at St Helier as GMT, results in times at Le Légué in French Standard Time (GMT + 0100) without further adjustment.

	Lat	Long	Time Difference HW		LW		Height Difference (metres)			
ST HELIER See p. 114	49°11′N	2°07′W	0100 and 1300	0800 and 2000	0200 and 1400	0700 and 1900	11.1	8.1	4.1	1.3
Le Légué	48°32′N	2°44′W	+0030	+0045	+0035	+0031	+0.3	+0.6	0.0	+0.1

TIDES relative to HW Dover (approx) HW −0520. Stream sets inwards +0130; outwards −0455.

GENERAL

Le Légué has two wet docks divided by a swing bridge and yachts use the innermost. St Brieuc, from which the Bay takes its name, is about a mile up the hill.

APPROACH

The approach is simple in offshore winds but dangerous in strong northerlies. The whole area dries but there is good holding about 2 miles NE of the harbour entrance in 3m. The channel dries 5.7m, has about 5.8m at MHWS and is buoyed. The lock gates open at times depending on the height of tide at St Malo as follows: 9m–10m: −1HW+1; 10m–11m: −1.1/4HW+1.1/4; 11m–11.5m: −1.1/2HW+1.1/2; above 11.5m: −2HW+1.1/2. There are waiting berths on either side below the lock which dry to about 6m. Limited facilities. There is an Aero Radio Beacon in position 48°34′.1N 2°46′.9W, 'SB', 353kHz, 25M.

CAUTIONS

1 Keep clear in strong northerlies.
2 A blue flag at the lock indicates that sluicing is imminent.
3 There are numerous shell fish beds in the bay.

TELEPHONES

Harbour Master:	(96) 33 35 41
Customs:	(96) 33 33 03
Medical:	(96) 61 49 07 (Doctor)
British Consul:	(99) 46 26 64

VHF

Le Légué Port: Ch 16; 12

YACHT CLUBS

La Toupie
Association Nautique du Légué
Centre social du Légué

The chart opposite is based on Chart No 3674 of the Service Hydrographique et Océanographique de la Marine (SHOM), reproduced by permission.

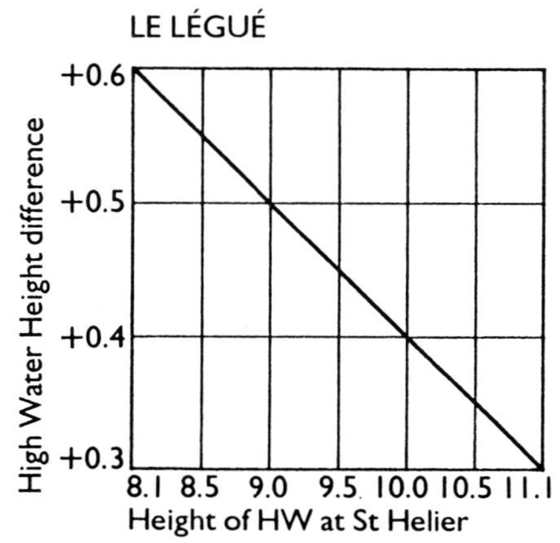

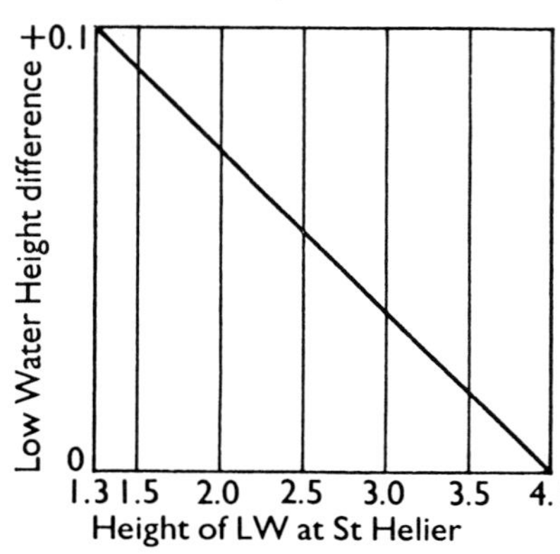

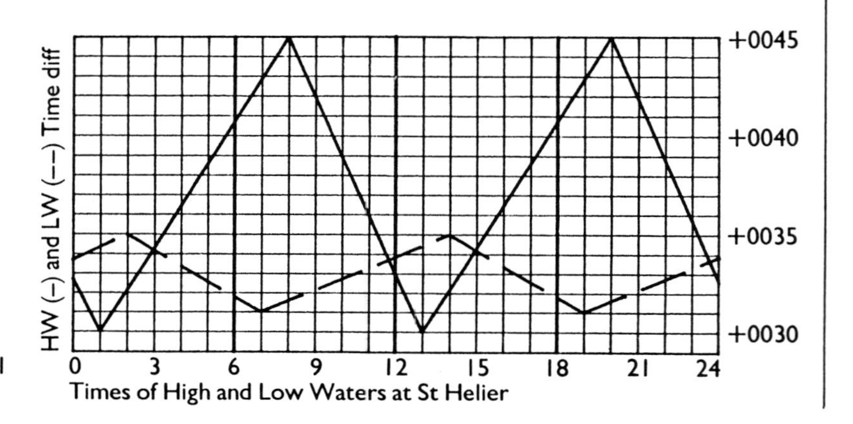

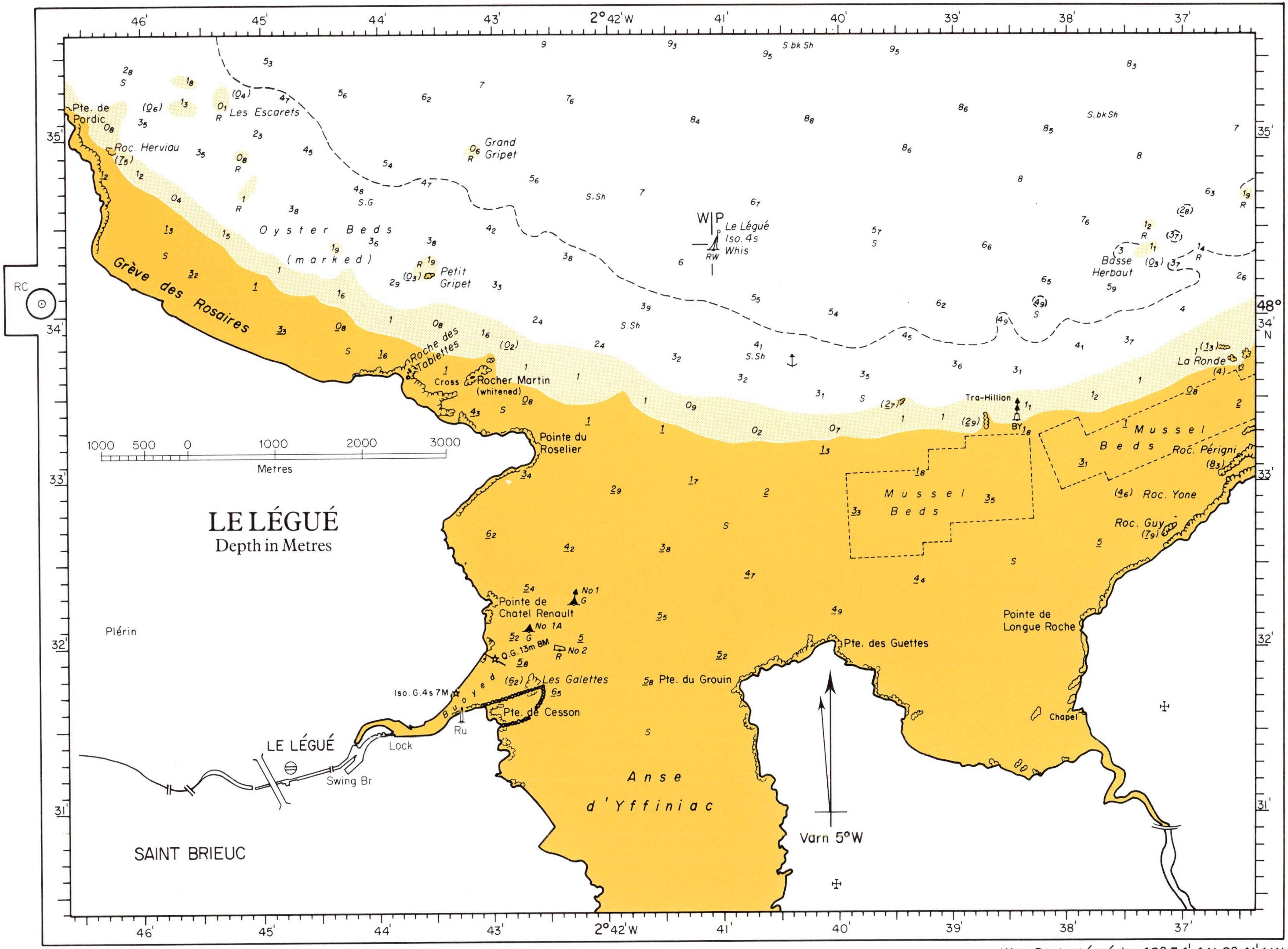

Chart No. 332

LE LÉGUÉ

Depth in Metres

Metres

1000 500 0 1000 2000 3000

Pte. de Pordic

Roc. Herviau

Les Escarets

Grand Gripet

Petit Gripet

Oyster Beds (marked)

Grève des Rosaires

Roche des Tablettes

Cross

Rocher Martin (whitened)

Pointe du Roselier

WP Le Légué Iso. 4s Whis

Tra-Hillion

Mussel Beds

Mussel Beds

Roc. Périgni

Roc. Yone

Roc. Guy

La Ronde

Basse Herbaut

Pointe de Longue Roche

Chapel

Pte. des Guettes

Varn 5°W

Pte. du Grouin

Anse d'Yffiniac

Pointe de Chatel Renault

No 1 G

No 1A

No 2 R

Q.G.13m 8M

Les Galettes

Pte. de Cesson

Iso. G.4s 7M

Buoyed

Ru

Lock

Swing Br

LE LÉGUÉ

Plérin

SAINT BRIEUC

RC

Way Pt: Le Légué by 48° 34'·4 N 2° 41'·1 W

Binic

See Chart No 333.

TIDE

Time and height differences on Standard Port (for instructions in use, see p. xii)
Note that this table, using times at St Helier as GMT, results in times at Binic in French Standard Time (GMT +0100) without further adjustment.

	Lat	Long	Time Difference HW		LW		Height Difference (metres)			
ST HELIER See p. 114	49°11′N	2°07′W	0100 and 1300	0800 and 2000	0200 and 1400	0700 and 1900	11.1	8.1	4.1	1.3
Binic	48°36′N	2°49′W	+0030	+0045	+0035	+0031	+0.3	+0.6	0.0	+0.1

TIDES relative to HW Dover (approx) HW−0520. Stream sets SE +0105; NW −0510. There is an eddy E along the S jetty between +0500 and +0600.

GENERAL

Binic has a drying outer harbour, with good drying berths and a locked inner harbour with pontoon berths for 500 plus 50 visitors' berths alongside the quay immediately to port. Most facilities in the season.

APPROACH

Apart from Basse Gouin (1.6m) about 1.1/2 miles ENE of the entrance, the approach is clear of dangers. The outer harbour dries up to 6m and is accessible −3HW+3. The lock opens, during working hours, when the height of water reaches 9.5m and closes at HW. There is an Aero Radio Beacon in position 48°34′.1N 2°46′.9W, 'SB', 353kHz, 25M.

CAUTION

1 A strong E wind makes access difficult.

TELEPHONES

Harbour Master:	(96) 73 61 86
Customs:	(96) 33 33 03
Medical:	(96) 42 61 05 (Doctor)
British Consul:	(99) 46 26 64

YACHT CLUB

Club Nautique du Binic: (96) 42 62 72

The chart opposite is based on Chart No 3674 of the Service Hydrographique et Océanographique de la Marine (SHOM), reproduced by permission.

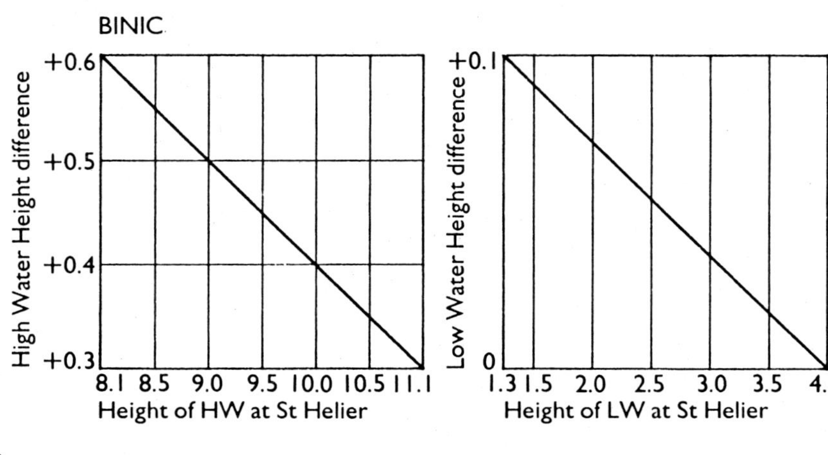

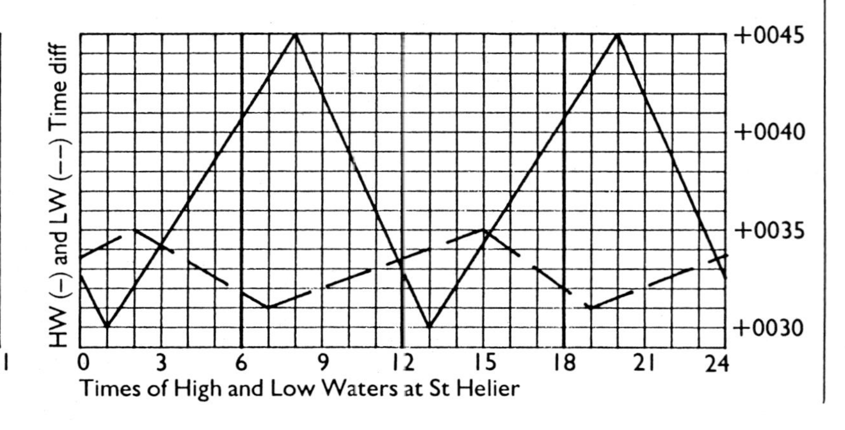

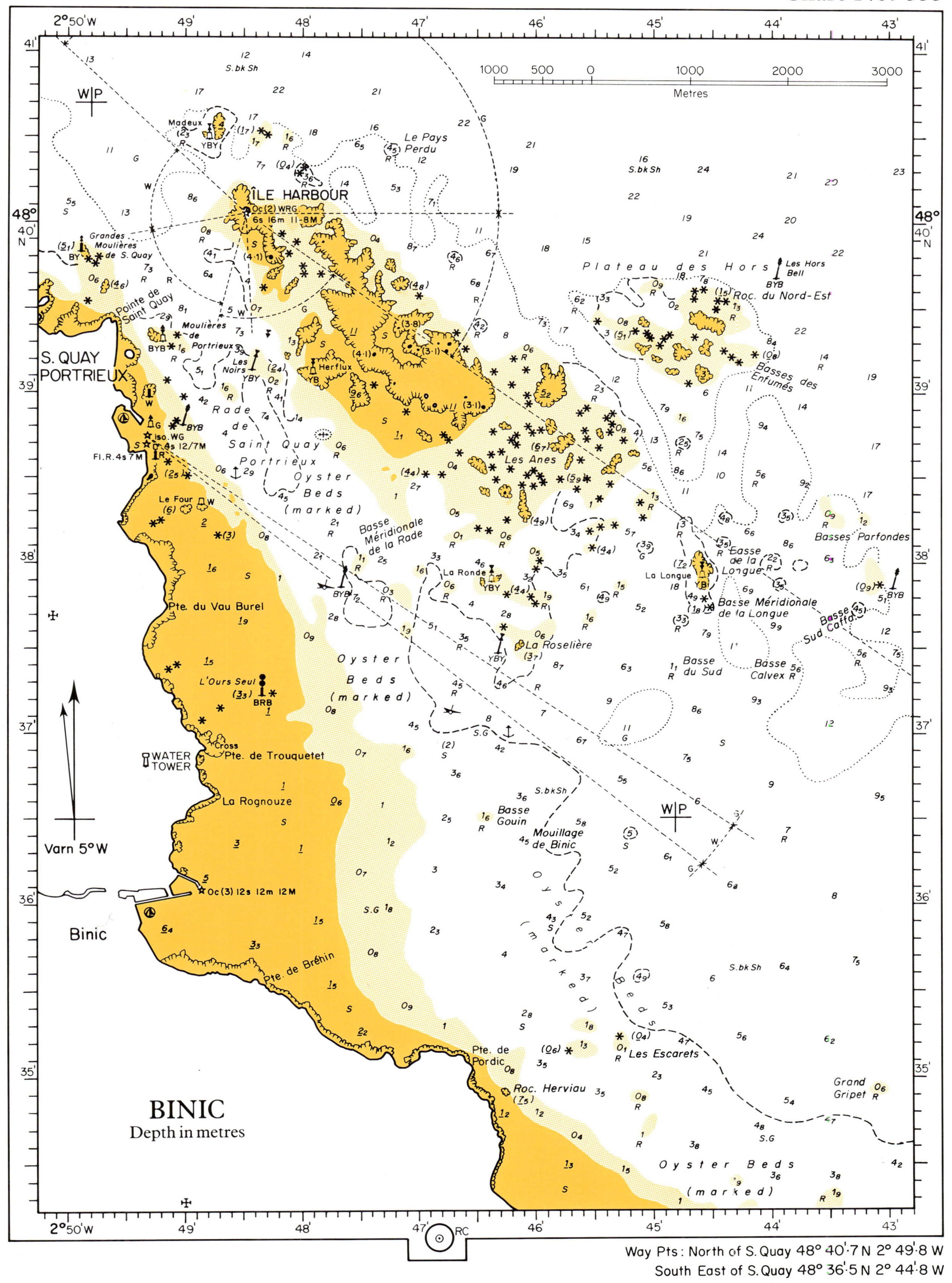

Way Pts: North of S. Quay 48° 40'·7 N 2° 49'·8 W
South East of S. Quay 48° 36'·5 N 2° 44'·8 W

St Quay-Portrieux

See Chart No 334.

TIDE

Time and height differences on Standard Port (for instructions in use, see p. xii)
Note that this table, using times at St Helier as GMT, results in times at Portrieux in French Standard Time (GMT +0100) without further adjustment.

	Lat	Long	Time Difference HW		LW		Height Difference (metres)			
ST HELIER See p. 114	49°11′N	2°07′W	0100 and 1300	0800 and 2000	0200 and 1400	0700 and 1900	11.1	8.1	4.1	1.3
Portrieux	48°38′N	2°49′W	+0030	+0045	+0030	+0030	+0.3	+0.5	0.0	+0.1

TIDES relative to HW Dover (approx) HW −0520. Stream sets SSE +0105; NW −0510.

GENERAL

St Quay-Portrieux is a drying harbour, well sheltered. There are plans for a marina to the N of the existing harbour. Stay is limited to three days in July and August. Some facilities.

APPROACH

For a general approach chart see Chart No. 333. The northern approach is easiest by day; the southern, lit, approach by night. The harbour dries 4m and yachts use the S part. Anchor off, if necessary in about 3m half a mile SE of the entrance.

CAUTION

1 The mole which forms the S arm of the harbour is formed by rocks and its sides are not vertical; keep well clear.

TELEPHONES

Harbour Master: (96) 70 52 04
Customs: (96) 33 33 03
Medical: (96) 70 41 31
British Consul: (99) 46 26 64

YACHT CLUB

Cercle de la Voile de Portrieux: (96) 70 41 76

The chart opposite is based on Chart No 3672 of the Service Hydrographique et Océanographique de la Marine (SHOM), reproduced by permission.

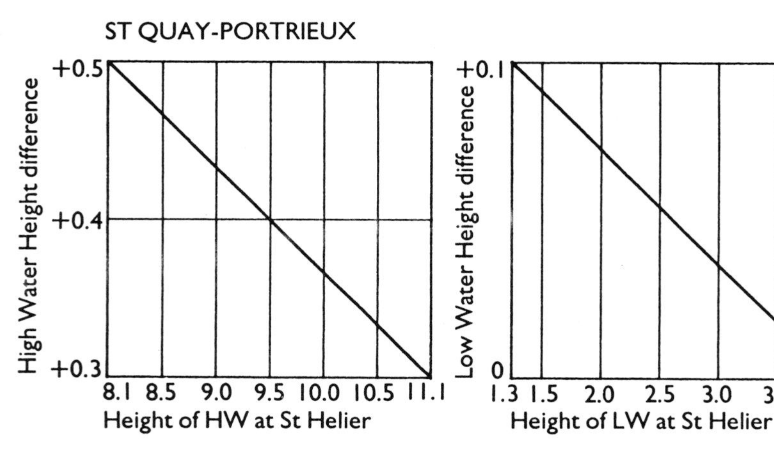

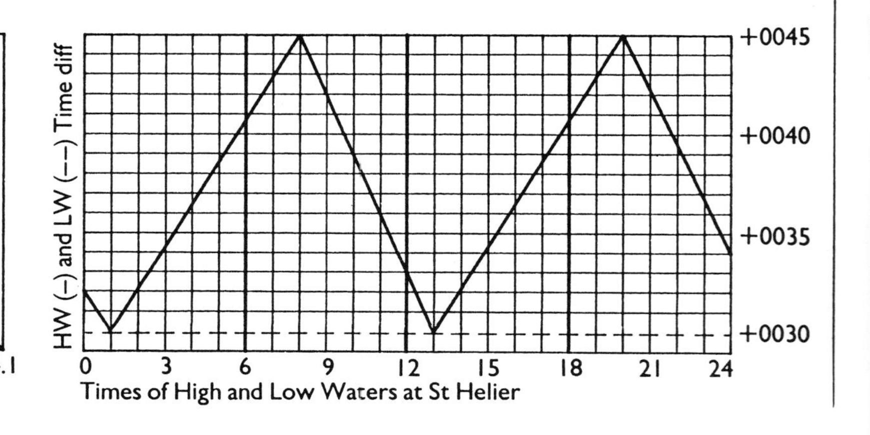

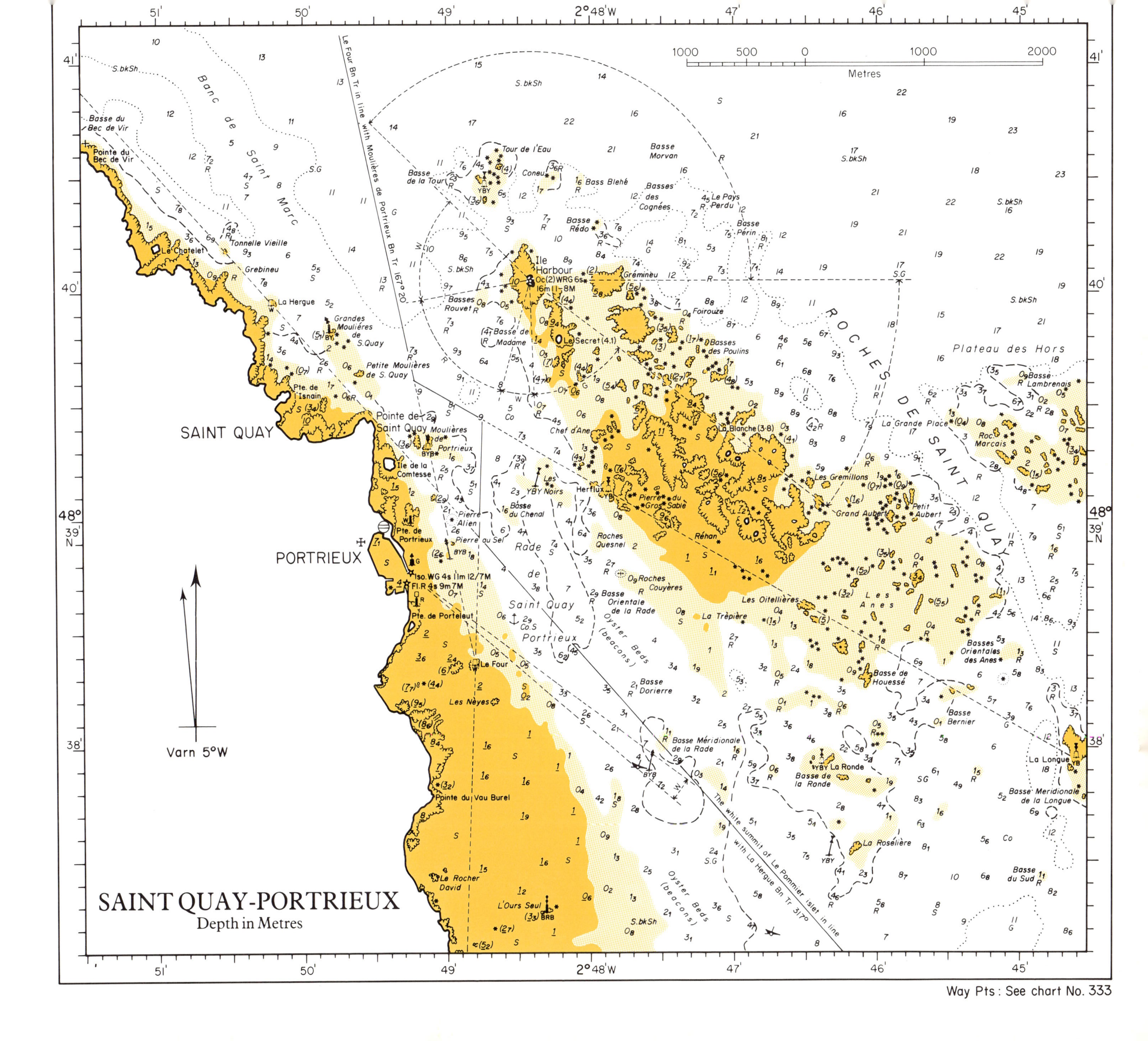

SAINT QUAY-PORTRIEUX
Depth in Metres
Way Pts: See chart No. 333
ROCHES DE SAINT QUAY
SAINT QUAY
PORTRIEUX
Varn 5°W
Metres
Plateau des Hors
Banc de Saint Marc
Rade de Saint Quay Portrieux
Le Four Bn Tr in line with Moulières de Portrieux Bn Tr 167°20'
The white summit of Le Pommier islet in line with La Hergue Bn Tr 317°
Oyster Beds (beacons)
Ile Harbour
Les Oitellières
Les Anes

Paimpol

See Chart No 335.

TIDE

Time and height differences on Standard Port (for instructions in use, see p. xii)
Note that this table, using times at St Helier as GMT, results in times at Paimpol in French Standard Time (GMT +0100) without further adjustment.

	Lat	Long	Time Difference HW		LW		Height Difference (metres)			
ST HELIER See p. 114	49°11′N	2°07′W	0100 and 1300	0800 and 2000	0200 and 1400	0700 and 1900	11.1	8.1	4.1	1.3
Paimpol	48°47′N	3°02′W	+0025	+0038	+0025	+0040	−0.6	−0.3	−0.9	−0.7

TIDES relative to HW Dover (approx) HW−0515. Stream (at Les Charpentiers) sets SE +0105; NW −0610.

GENERAL

Paimpol is a small commercial and fishing port with good accommodation and shelter for yachts. The approach dries completely. Most facilities.

APPROACH

See the chart for leading lines and lights. Note the marked anchorages. The approach, which dries up to 4.9m, is marked with unlit buoys and beacons and care should be taken to avoid the many rocks, especially to the N of the channel. The lock usually opens for yachts 1½ hours either side of HW. Visiting yachts berth in the first basin (No 2) where there is between 3.0 and 4.6m depending on the range of tide.

CAUTIONS

1 When both lock gates are open, the flood may give up to 2 knots through the lock.
2 See notes above on approach dangers.
3 There are numerous oyster beds on both sides of the approach.

TELEPHONES

Harbour Master:	(96) 20 84 30
Customs:	(96) 20 95 38
Medical:	(96) 20 80 04
British Consul:	(99) 46 26 64

MARINA

Port de Plaisence:	(96) 20 80 15

YACHT CLUB

Centre Nautique des Glenans:	(96) 20 84 33

The chart opposite is based on Chart No 3673 of the Service Hydrographique et Océanographique de la Marine (SHOM), reproduced by permission.

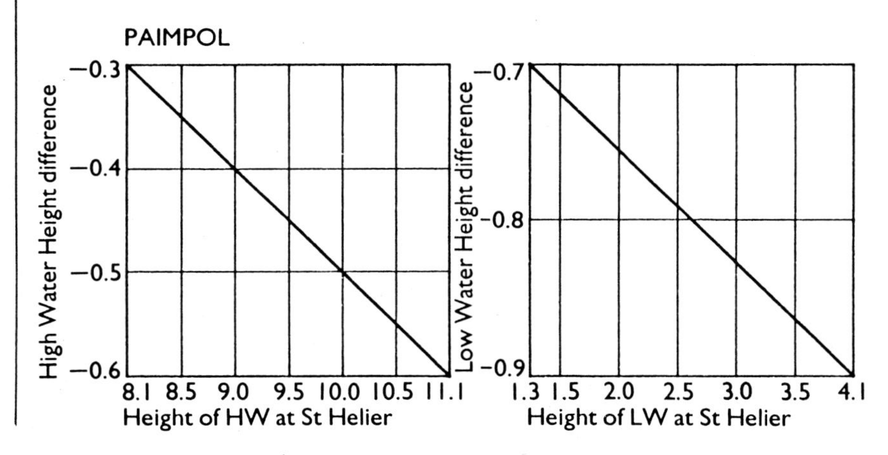

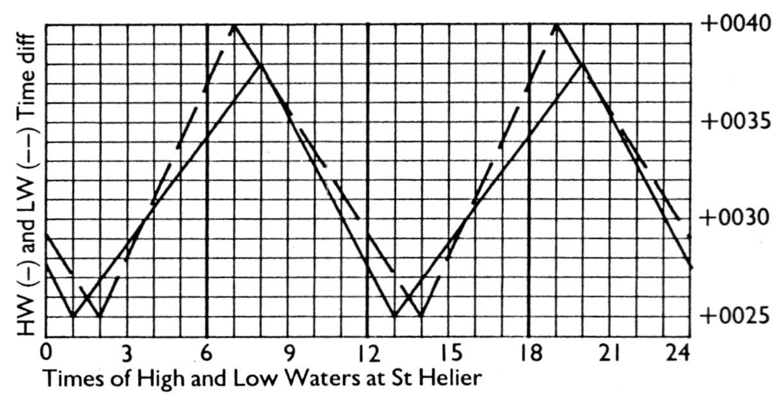

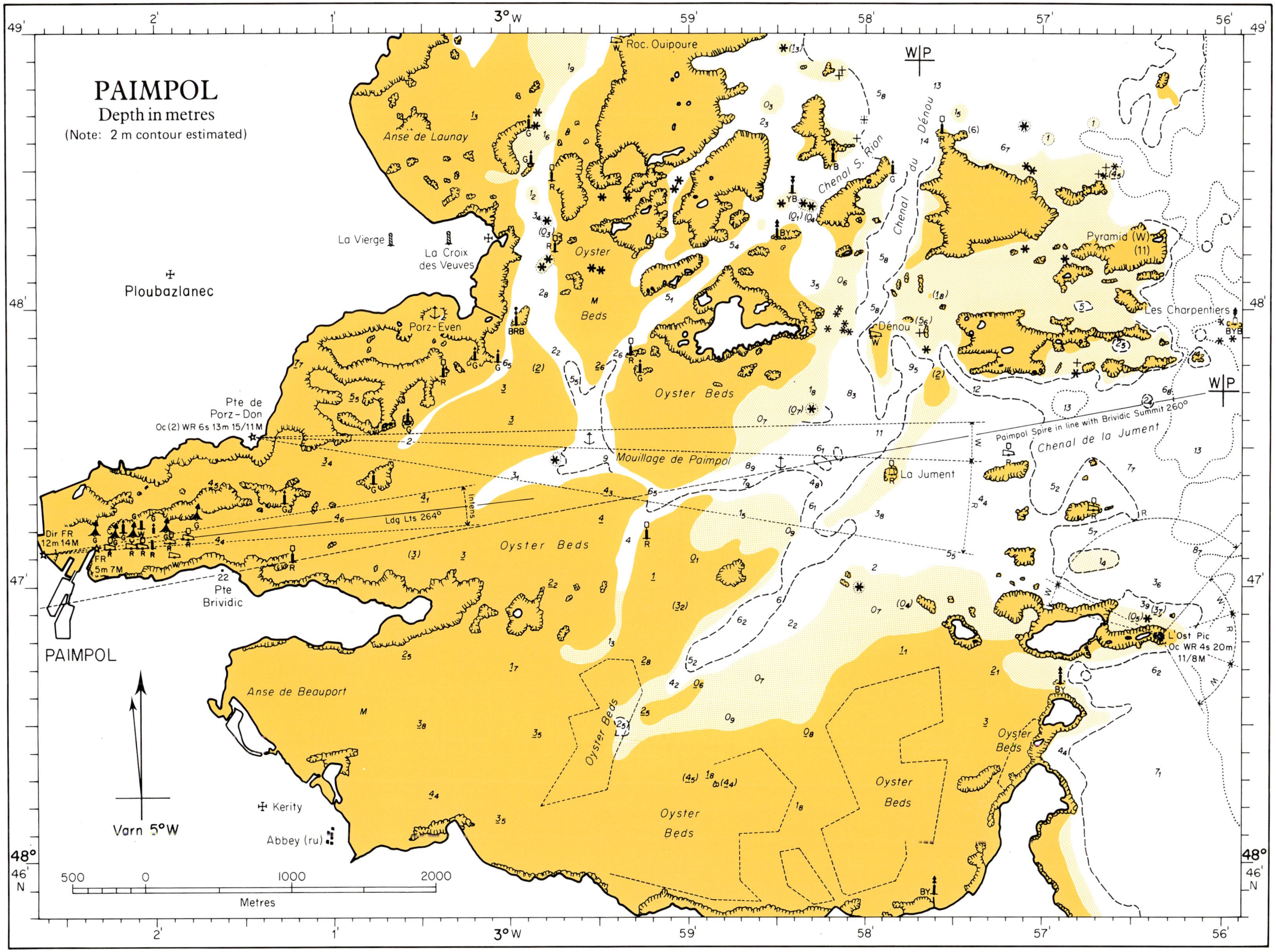

PAIMPOL

Depth in metres

(Note: 2 m contour estimated)

Ploubazlanec
La Vierge
La Croix des Veuves
Anse de Launay
Porz-Even
Pte de Porz-Don
Oc(2) WR 6s 13m 15/11M
Dir FR 12m 14M
FR 5m 7M
Pte Brividic
PAIMPOL
Anse de Beauport
Kerity
Abbey (ru)
Roc. Ouipoure
Oyster Beds
Mouillage de Paimpol
Ldg Lts 264°
Intens
Chenal S. Rion
Chenal du Dénou
Dénou
La Jument
Chenal de la Jument
Paimpol Spire in line with Brividic Summit 260°
Pyramid (W) (11)
Les Charpentiers
L'Ost Pic
Oc WR 4s 20m 11/8M
W|P
Varn 5°W
500 0 1000 2000
Metres

Way Pts: Chenal de la Jument 48°47'·7N 2°56'·0W
Chenal du Dénou 48°48'·9N 2°57'·7W

Île de Bréhat

See Chart No 336.

TIDE

Time and height differences on Standard Port (for instructions in use, see p. xii)
Note that this table, using times at St Helier as GMT, results in times at Bréhat in French Standard Time (GMT +0100) without further adjustment.

	Lat	Long	Time Difference HW		LW		Height Difference (metres)			
			0100 and 1300	0800 and 2000	0200 and 1400	0700 and 1900				
ST HELIER See p. 114	49°11′N	2°07′W					11.1	8.1	4.1	1.3
Île de Bréhat	48°51′N	3°00′W	+0020	+0040	+0010	+0015	−0.6	−0.1	−0.4	−0.1

TIDES relative to HW Dover (approx) HW (Les Heaux) −0520. Stream sets (Les Heaux) SE +0135; NW −0440: (close E and W of the island) S +0105; N −0510.

GENERAL

Île de Bréhat is a popular holiday centre but has almost no facilities for visiting craft. There are no permanent alongside berths but plenty of room to anchor in shelter.

APPROACH

Entry from the E is shown on the chart opposite and from the N on the next chart. The former is lit only by one light and is best used by day when it is well marked and perfectly safe, though a little daunting for the first time. Anchor anywhere in the Rade, provided that the weather is reasonable, but not within the cable area which is marked on the chart and, though not completely, by beacons. At neaps, the other anchorages provide more interest and better shelter, though all of them need care due to existing moorings. Port Clos allows a very temporary alongside berth when the tide serves and the ferry is not there. There is a Radio Beacon on Rosedo lighthouse, 'DO', 294.2kHz, 10M.

CAUTIONS

1 Streams run strongly.
2 Note the cable area across the Rade and at the head of La Chambre where anchoring is forbidden.
3 There are many oyster beds in the area.

TELEPHONES

See Paimpol or Lézardrieux on pages 70 and 76.

YACHT CLUB

Club Nautique
de Bréhat: (96) 20 00 69

The chart opposite is based on Chart No 3673 of the Service Hydrographique et Océanographique de la Marine (SHOM), reproduced by permission.

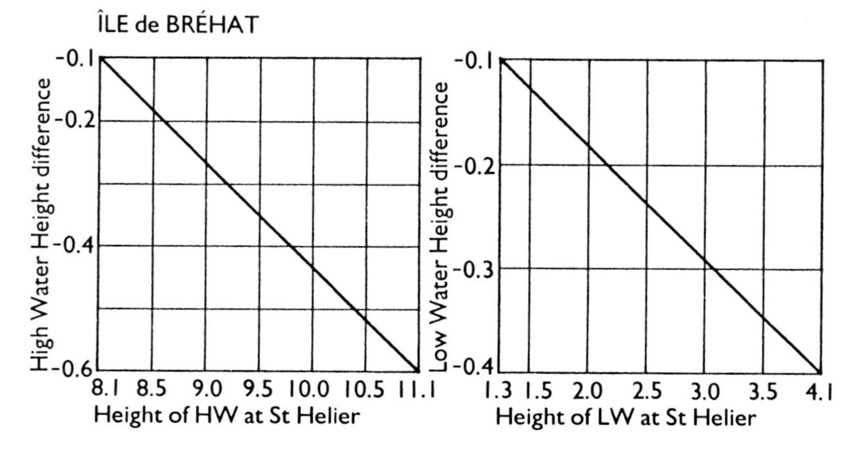

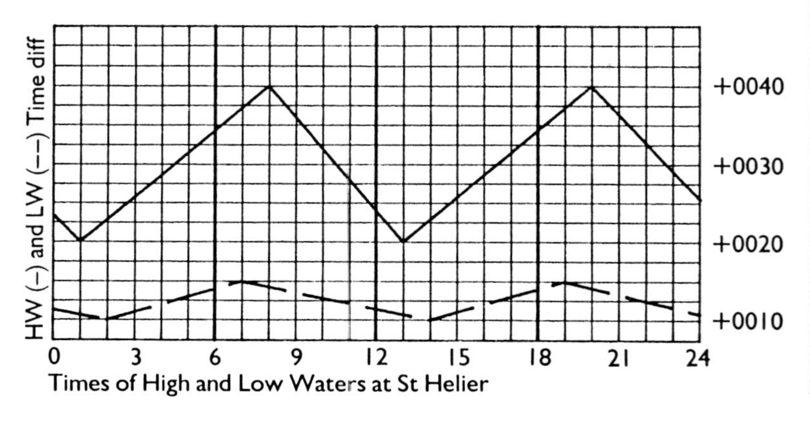

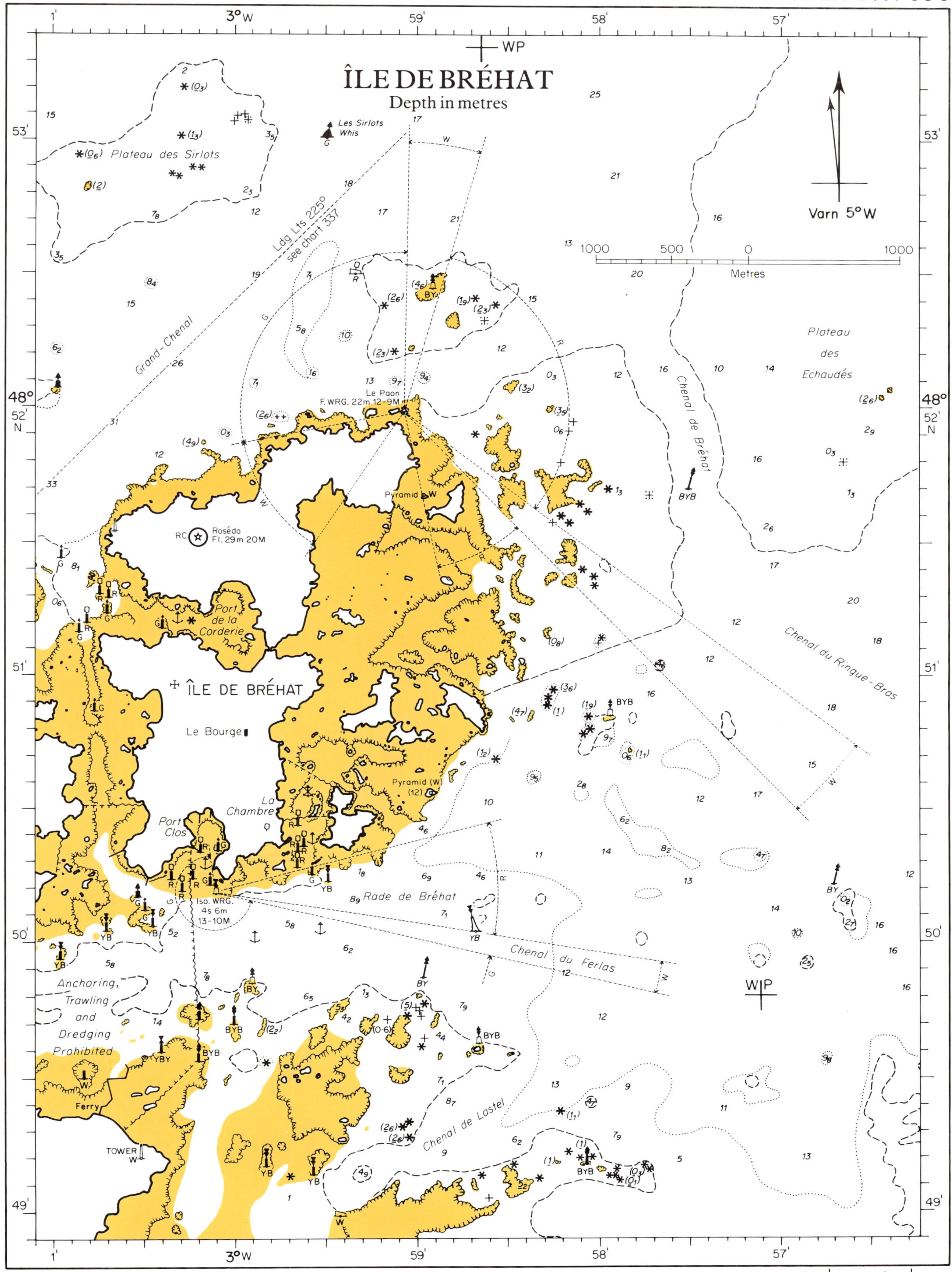

Way Pts: Grand-Chenal 48° 53'·35 N 2° 58'·65 W
Chenal du Ferlas 48° 49'·8 N 2° 57'·1 W

Approaches to Rivière de Trieux

See Chart No 337.

See also page 72.

TIDE

See Île de Bréhat (page 72) and Lézardrieux (page 76).

GENERAL

This chart covers the N approach to the Rivière de Trieux and Lézardrieux and the W side of Île de Bréhat.

APPROACH

The best approach, without previous experience of the area, is the Grand Chenal which is well lit and marked. There are anchorages shown on the previous chart and on the chart opposite, especially just NE of the small drying harbour of Loguivy, which, itself, is not recommended for yachts.

CAUTIONS

1 The streams run strongly.
2 There are many oyster beds in the area.

TELEPHONES

See Lézardrieux on page 76.

The chart opposite is based on Chart No 3673 of the Service Hydrographique et Océanographique de la Marine (SHOM), reproduced by permission.

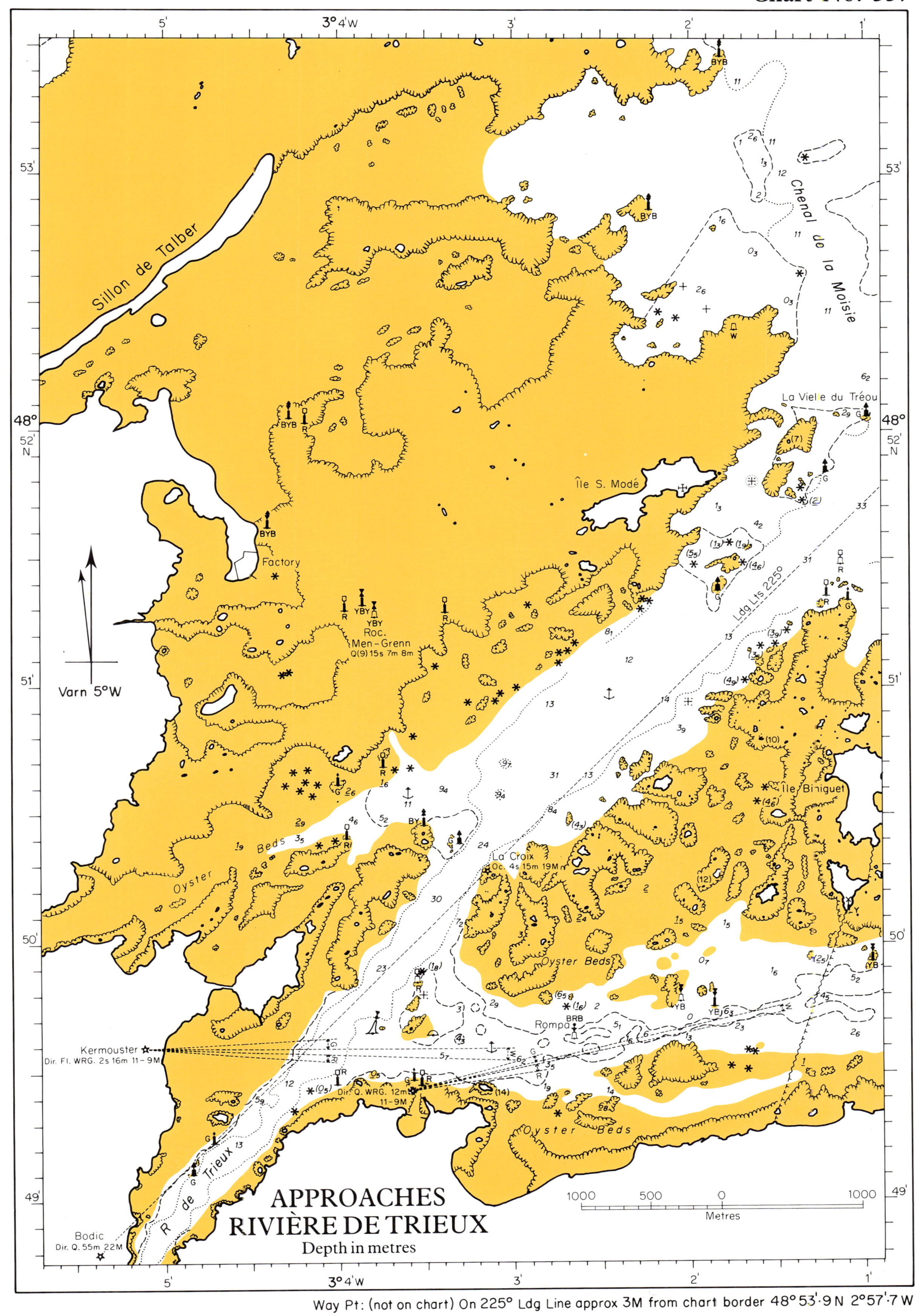

Way Pt: (not on chart) On 225° Ldg Line approx 3M from chart border 48°53'·9 N 2°57'·7 W

Rivière de Trieux and Lézardrieux

See Chart No 338.

TIDE

Time and height differences on Standard Port (for instructions in use, see p. xii)
Note that this table, using times at St Helier as GMT, results in times at Lézardrieux in French Standard Time (GMT +0100) without further adjustment.

	Lat	Long	Time Difference HW		LW		Height Difference (metres)			
			0100 and 1300	0800 and 2000	0200 and 1400	0700 and 1900				
ST HELIER See p. 114	49°11′N	2°07′W					11.1	8.1	4.1	1.3
Lézardrieux	48°47′N	3°06′W	+0026	+0038	+0015	+0035	−0.9	−0.5	−0.6	−0.3

GENERAL

Lézardrieux lies on the W bank of the Pontrieux River in perfect shelter and with a marina and moorings off. Most facilities. Above Lézardrieux there is a suspension bridge, 17.7m clearance, and then, after about 6 miles, the town of Pontrieux which has a wet dock, 3.9m, formed by the river, the gates of which open −1HW+1.

APPROACH

See page 74.

CAUTION

1 The streams run strongly.

TELEPHONES

Harbour Master:	(96) 20 14 22
Customs:	See Paimpol on page 70
Medical:	(96) 20 10 30 (Doctor)
British Consul:	(99) 46 26 64

VHF

Lézardrieux Port:	Ch 9 (office hours)

MARINA

Lézardrieux:	(96) 20 14 22

YACHT CLUB

YC de Trieux:	(96) 20 10 39

The chart opposite is based on Chart No 3673 of the Service Hydrographique et Océanographique de la Marine (SHOM), reproduced by permission.

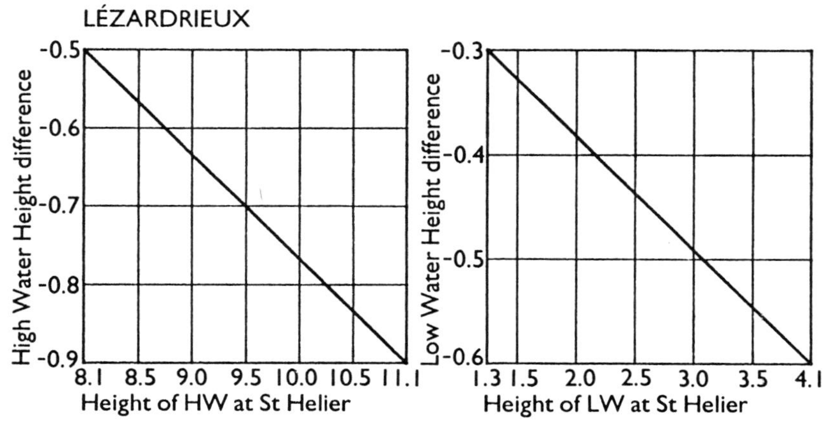

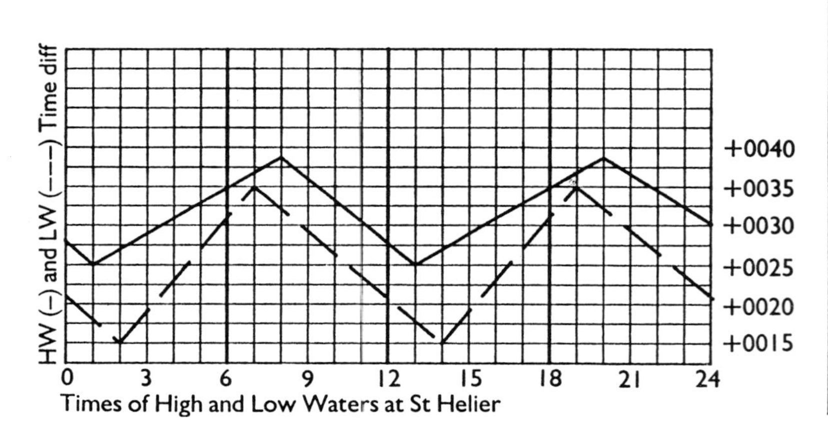

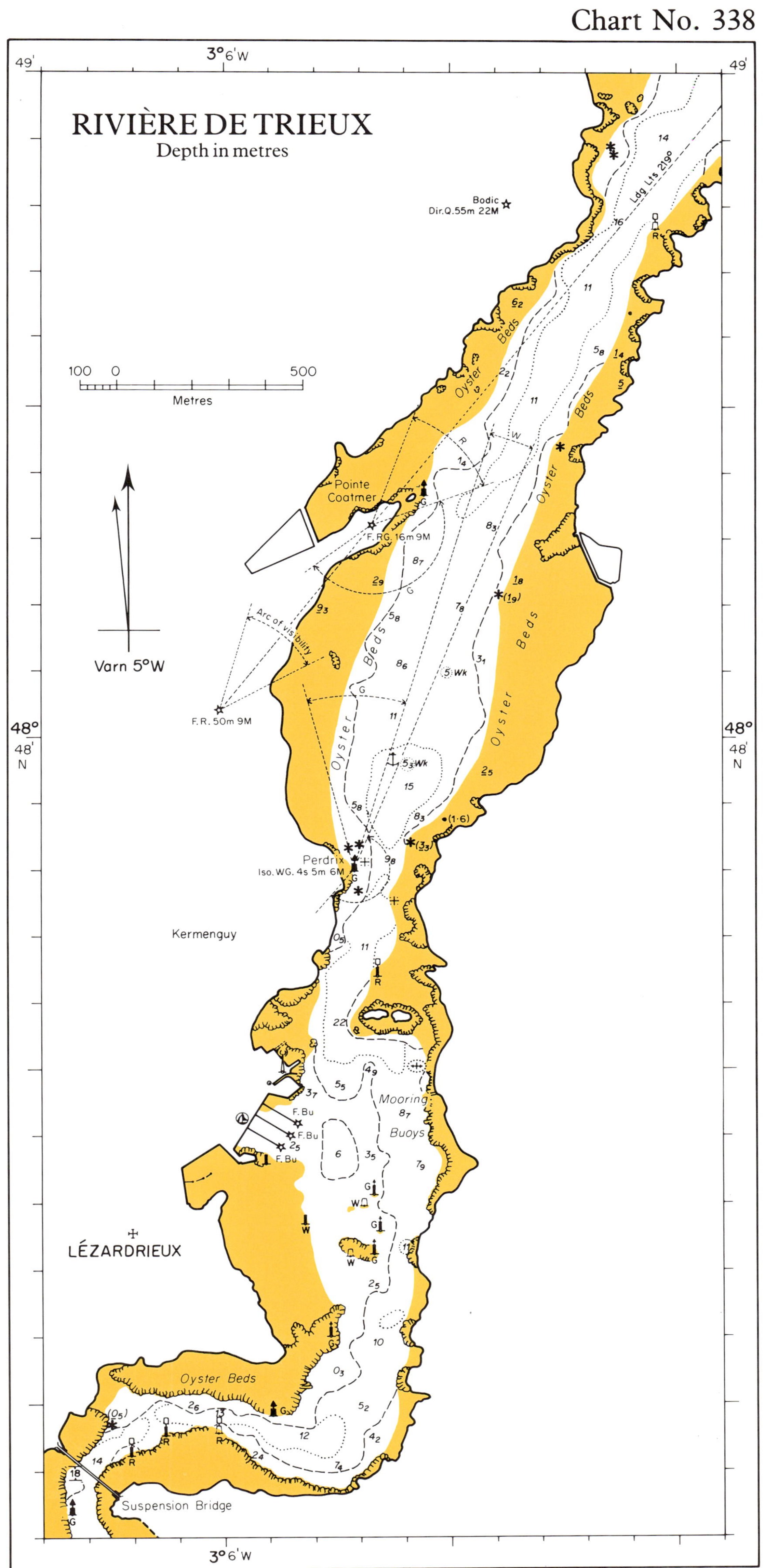
RIVIÈRE DE TRIEUX
Depth in metres
3°6'W
49'
48°
48'
N
100 0 500
Metres
Varn 5°W
Bodic
Dir.Q.55m 22M
Ldg Lts 219°
Oyster Beds
Pointe
Coatmer
F.RG. 16m 9M
Arc of visibility
F.R. 50m 9M
Perdrix
Iso.WG. 4s 5m 6M
Kermenguy
Mooring
Buoys
F.Bu
LÉZARDRIEUX
Suspension Bridge

Approaches to Rivière de Tréguier

See Chart No 339.

TIDE

Time and height differences on Standard Port (for instructions in use, see p. xii)
Note that this table, using times at St Helier as GMT, results in times at Les Heaux de Bréhat in French Standard Time (GMT +0100) without further adjustment.

	Lat	Long	Time Difference HW		LW		Height Difference (metres)			
ST HELIER See p. 114	49°11′N	2°07′W	0100 and 1300	0800 and 2000	0200 and 1400	0700 and 1900	11.1	8.1	4.1	1.3
Les Heaux de Bréhat	48°55′N	3°05′W	+0031	+0030	−0011	+0042	−1.2	−0.5	−0.7	−0.2

TIDES relative to HW Dover (approx) HW (Les Heaux) −0520. Stream sets between Les Heaux and the mainland ENE +0020; WSW +0445.

GENERAL

The chart and this information is provided as an aid for the mariner to reach the river and town of Tréguier.

APPROACH

There are three approaches to the R Tréguier. The Grande Passe is the only one lit and has plenty of water except for a kink in the channel at Pierre à l'Anglais, which dries 2.4m to the S of the leading line, and Basse du Corbeau with 4.4m actually on the leading line. The other two approaches require careful tidal calculations and clear weather.

CAUTIONS

1 The streams run strongly.
2 There are many oyster beds in the area.

TELEPHONES

See page 80.

The chart opposite is based on Chart No 3670 of the Service Hydrographique et Océanographique de la Marine (SHOM), reproduced by permission.

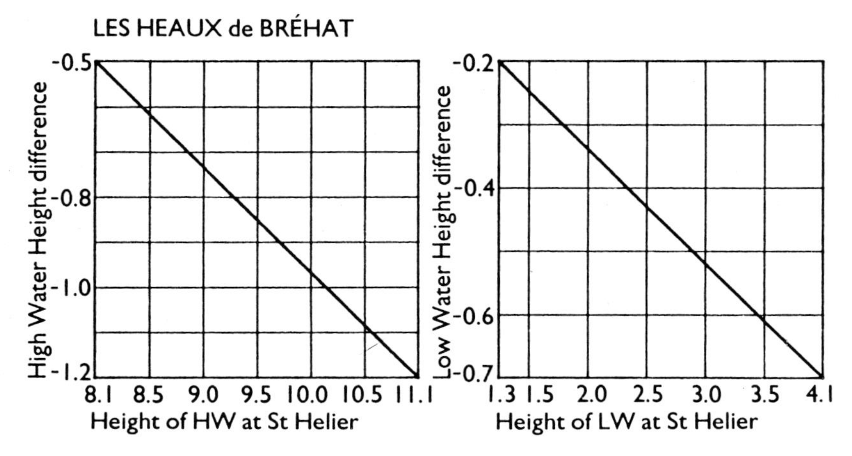

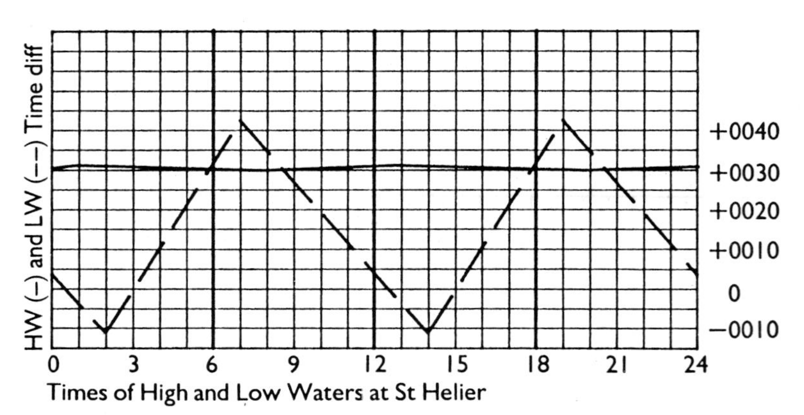

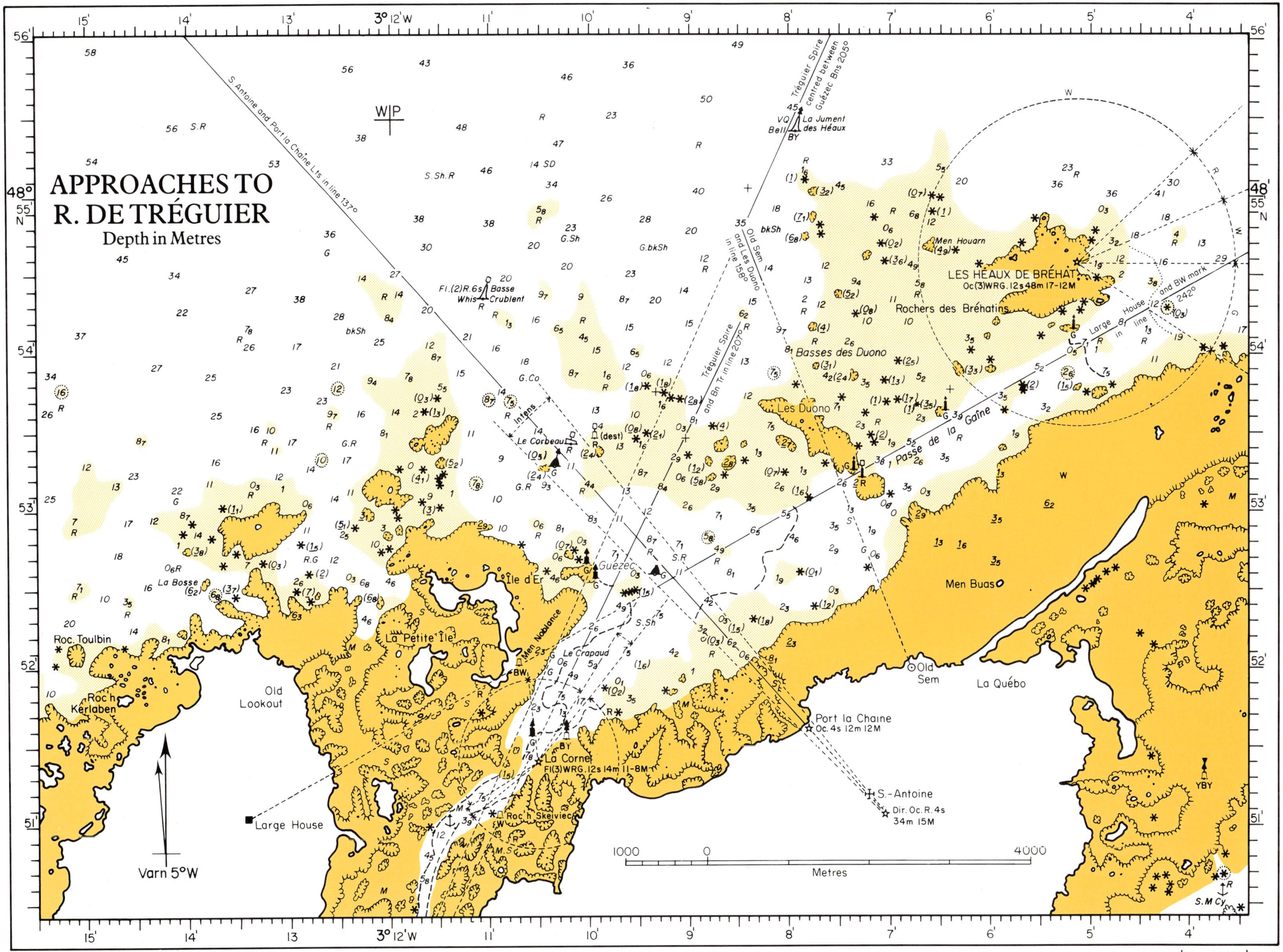

APPROACHES TO R. DE TRÉGUIER

Depth in Metres

Way Pt: 48°55'·5 N 3°12'·0 W

Tréguier

See Chart No 340.

TIDE

Time and height differences on Standard Port (for instructions in use, see p. xii)
Note that this table, using times at St Helier as GMT, results in times at Tréguier in French Standard Time (GMT +0100) without further adjustment.

	Lat	Long	Time Difference HW		LW		Height Difference (metres)			
ST HELIER See p. 114	49°11′N	2°07′W	0100 and 1300	0800 and 2000	0200 and 1400	0700 and 1900	11.1	8.1	4.1	1.3
Tréguier	48°47′N	3°13′W	−0004	+0012	−0025	+0015	−1.3	−0.6	−0.8	−0.2

TIDES relative to HW Dover (approx). Stream ebbs −0545; floods +0045.

GENERAL

There is excellent shelter and a marina with 130 visitors' berths, maximum length 12m. Most services.

APPROACH

See page 78. The river is navigable at any state of the tide but it is convenient to go up on the flood, just after LW when the banks are clearly visible. There are anchorages as shown on the chart but care must be taken to avoid oyster beds.

CAUTIONS

1 The stream runs very strong, through the marina.
2 There are many oyster beds in the area.

TELEPHONES

Harbour Master: (96) 92 42 37
Customs: (96) 92 31 44
Medical: (96) 92 32 14
British Consul: (99) 46 26 64

MARINA

Tréguier: (96) 92 30 19

YACHT CLUB

Club Nautique du Trégor: (96) 92 42 08

The chart opposite is based on Chart No 3672 of the Service Hydrographique et Océanographique de la Marine (SHOM), reproduced by permission.

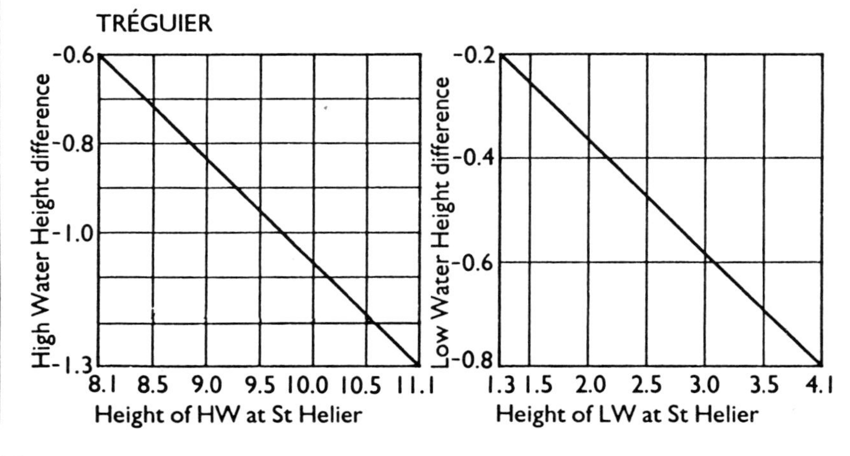

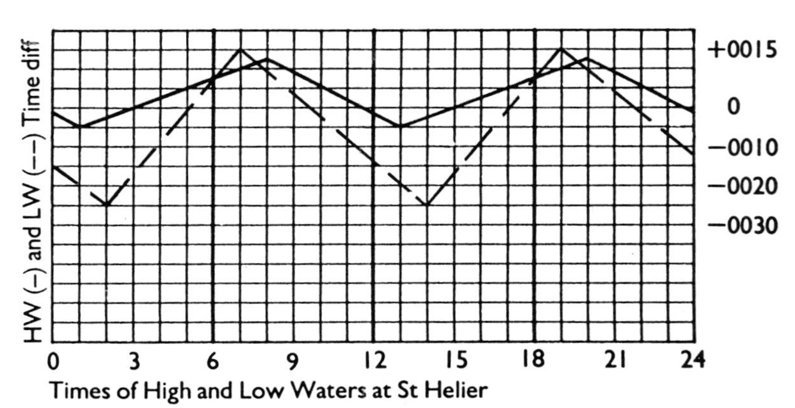

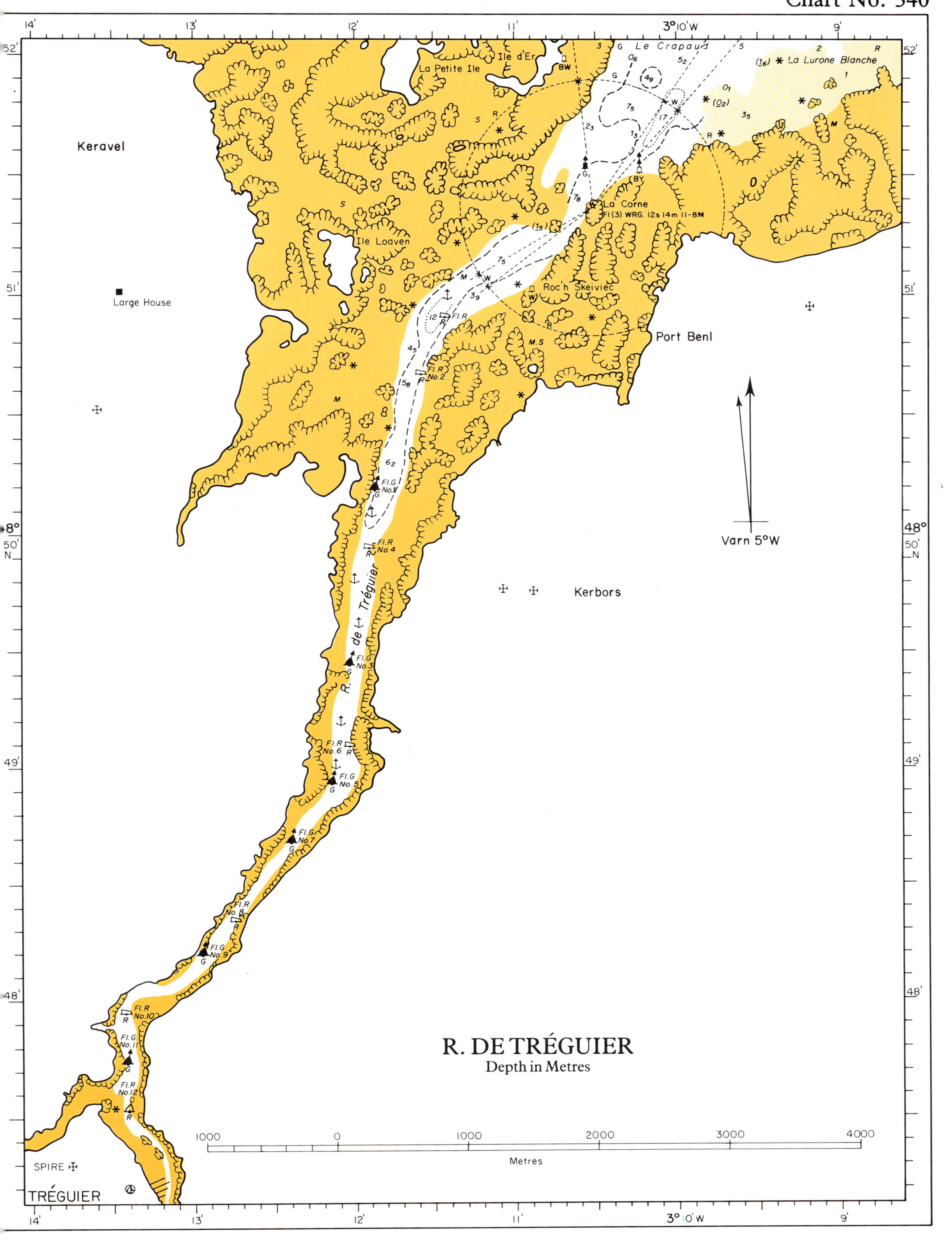

R. DE TRÉGUIER

Depth in Metres

Port Blanc

See Chart No 341.

TIDE

Time and height differences (estimated) on Standard Port (for instructions in use, see p. xii)
Note that this table, using times at St Helier as GMT, results in times at Port Blanc in French Standard Time (GMT +0100) without further adjustment.

	Lat	Long	Time Difference HW		LW		Height Difference (metres)			
			0100 and 1300	0800 and 2000	0200 and 1400	0700 and 1900				
ST HELIER See p. 114	49°11′N	2°07′W					11.1	8.1	4.1	1.3
Port Blanc	48°51′N	3°19′W	−0003	+0004	−0012	+0007	−1.6	−0.8	−0.6	0.0

TIDES relative to HW Dover (approx) HW −0555 Stream sets E +0020; W −0555

GENERAL

In settled weather or off-shore winds, Port Blanc provides a sheltered, drying harbour and an anchorage off, easy of access. Few facilities.

APPROACH

The approach is easy with a sectored light taking one in straight to the anchorage. The harbour dries and it may be possible to lie against a wall.

CAUTIONS

1 Onshore winds, especially with an ebb tide, can cause rough seas, making the anchorage dangerous.
2 There are rocks round the inside end of the northern breakwater.

The chart opposite is based on Chart No 3670 of the Service Hydrographique et Océanographique de la Marine (SHOM), reproduced by permission.

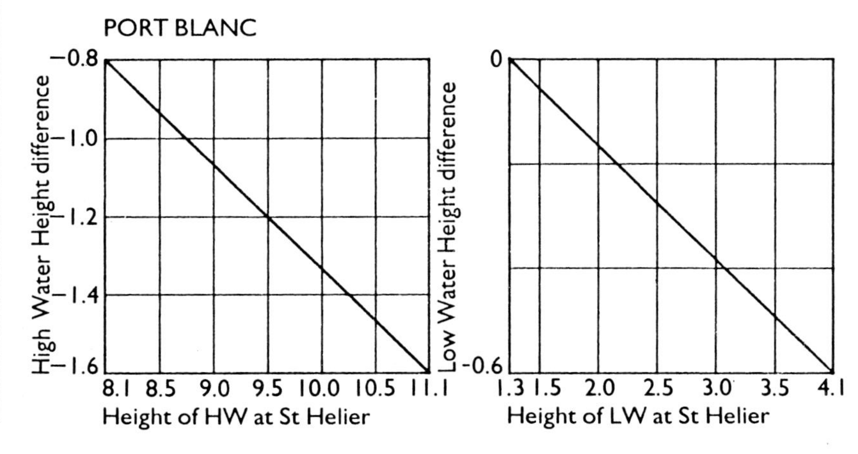

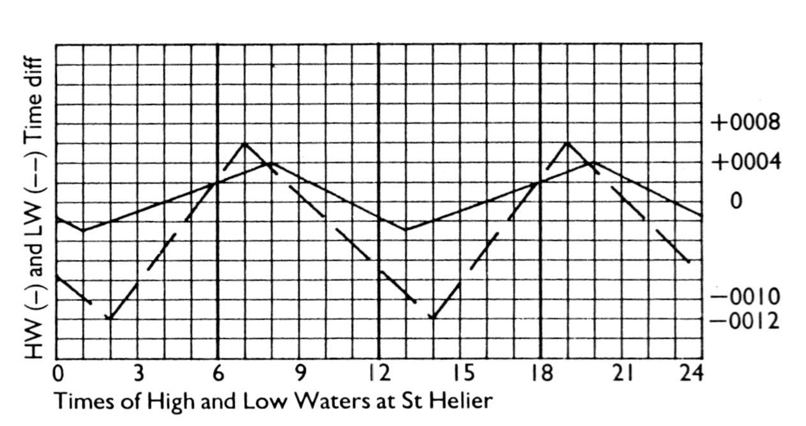

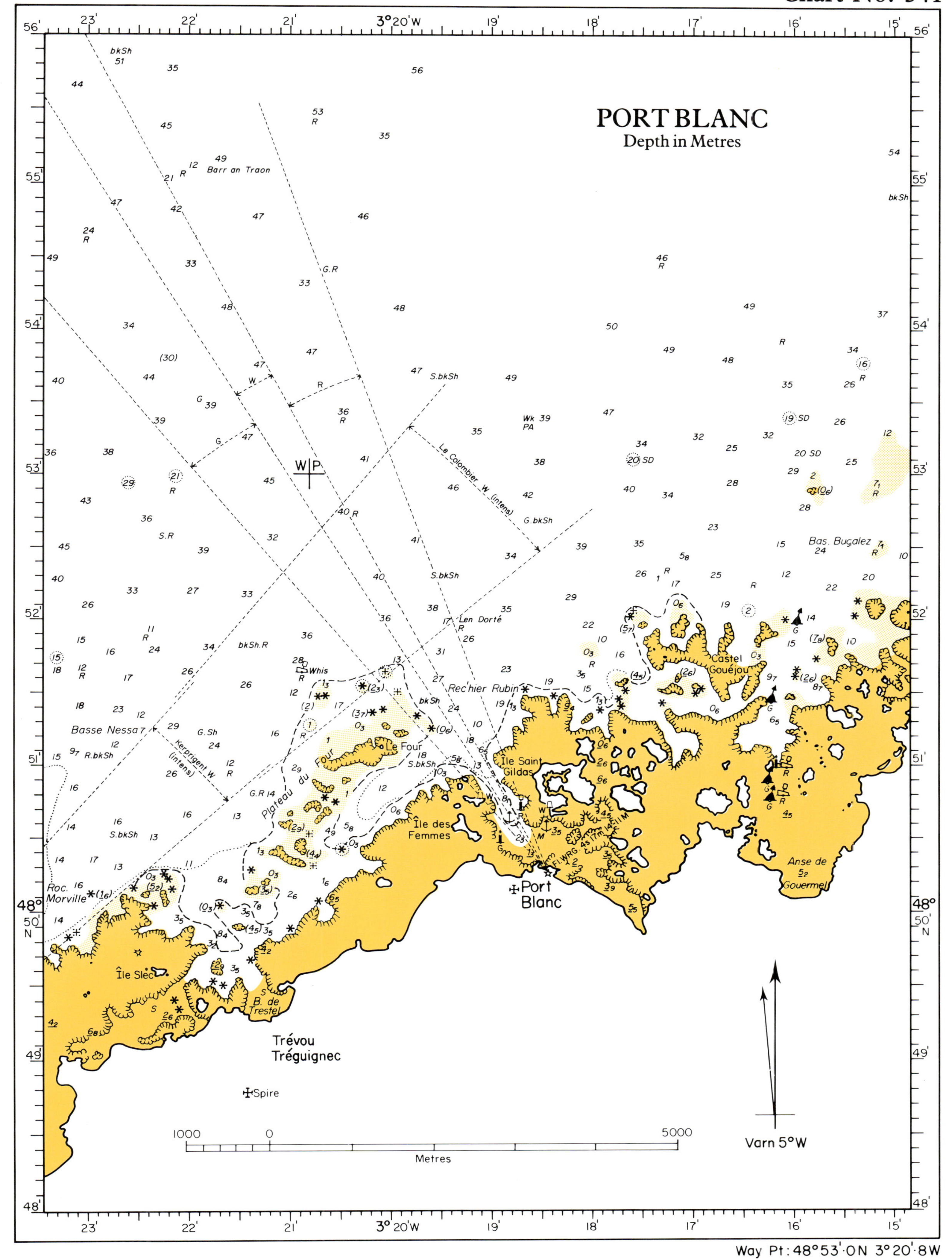
PORT BLANC
Depth in Metres
Barr an Traon
Le Colombier W (intens)
Basse Nessa
Kerprigent W (intens)
Plateau du Four
Le Four
Len Dorté
Rechier Rubin
Ile Saint Gildas
Île des Femmes
Port Blanc
Castel Gouéjou
Bas. Buçalez
Anse de Gouermel
Roc. Morville
Île Slec
B. de Trestel
Trévou Tréguignec
Spire
Metres
Varn 5°W
Way Pt: 48°53'·0N 3°20'·8W

Perros-Guirec

see Chart No 342.

For approaches, see chart on page 87.

TIDE

Time and height differences (estimated) on Standard Port (for instructions in use, see p. xii)
Note that this table, using times at St Helier as GMT, results in times at Perros-Guirec in French Standard Time (GMT +0100) without further adjustment.

	Lat	Long	Time Difference HW		LW		Height Difference (metres)			
ST HELIER See p. 28	49°11′N	2°07′W	0100 and 1300	0800 and 2000	0200 and 1400	0700 and 1900	11.1	8.1	4.1	1.3
Perros-Guirec	48°49′N	3°25′W	−0002	+0005	−0019	+0003	−1.9	−0.8	−0.7	−0.1

TIDES relative to HW Dover (approx) HW −0555.

Stream sets (off Les Couillons de Tomé) E +0035; W −0425.
(in Passe de l'Ouest) SE +0035; NW −0425.
(in Passe de l'Est) ENE +0035; WSW −0425.

GENERAL

Perros-Guirec offers excellent shelter and a good anchorage off. The marina has 600 berths, maximum 15m LOA, with 50 reserved for visitors. Most services.

APPROACH

See page 86 and Chart No. 343. There are two approaches, E and W, both lit. Both have isolated rocks very near the leading lines, so care is needed. The lock opens between −2HW+1 at springs and −1HW approaching neaps, though it is possible to be locked in for up to three days at neaps. There is at least 2m in the marina.

CAUTION

1 See Approach notes above.

TELEPHONES

Harbour Master:	(96) 23 37 82
Lock Master:	(96) 29 19 03
Customs:	(96) 23 18 12
Medical:	(96) 23 20 01
British Consul:	(99) 46 26 64

VHF

Perros-Guirec: Ch 16; 9

MARINA

Perros-Guirec: (96) 23 19 03

YACHT CLUB

S des Régates Perrosienne

The chart opposite is based on Chart No 3672 of the Service Hydrographique et Océanographique de la Marine (SHOM), reproduced by permission.

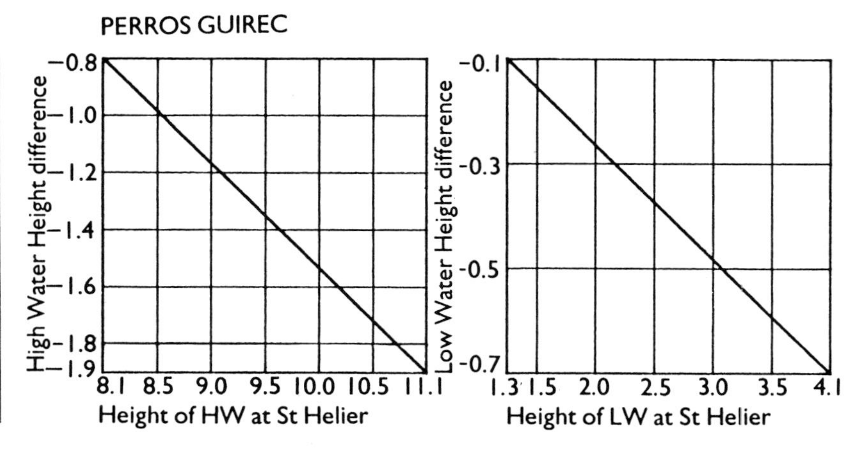

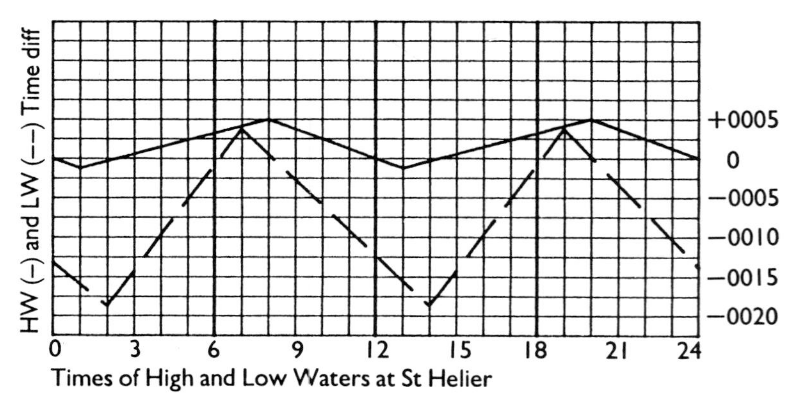

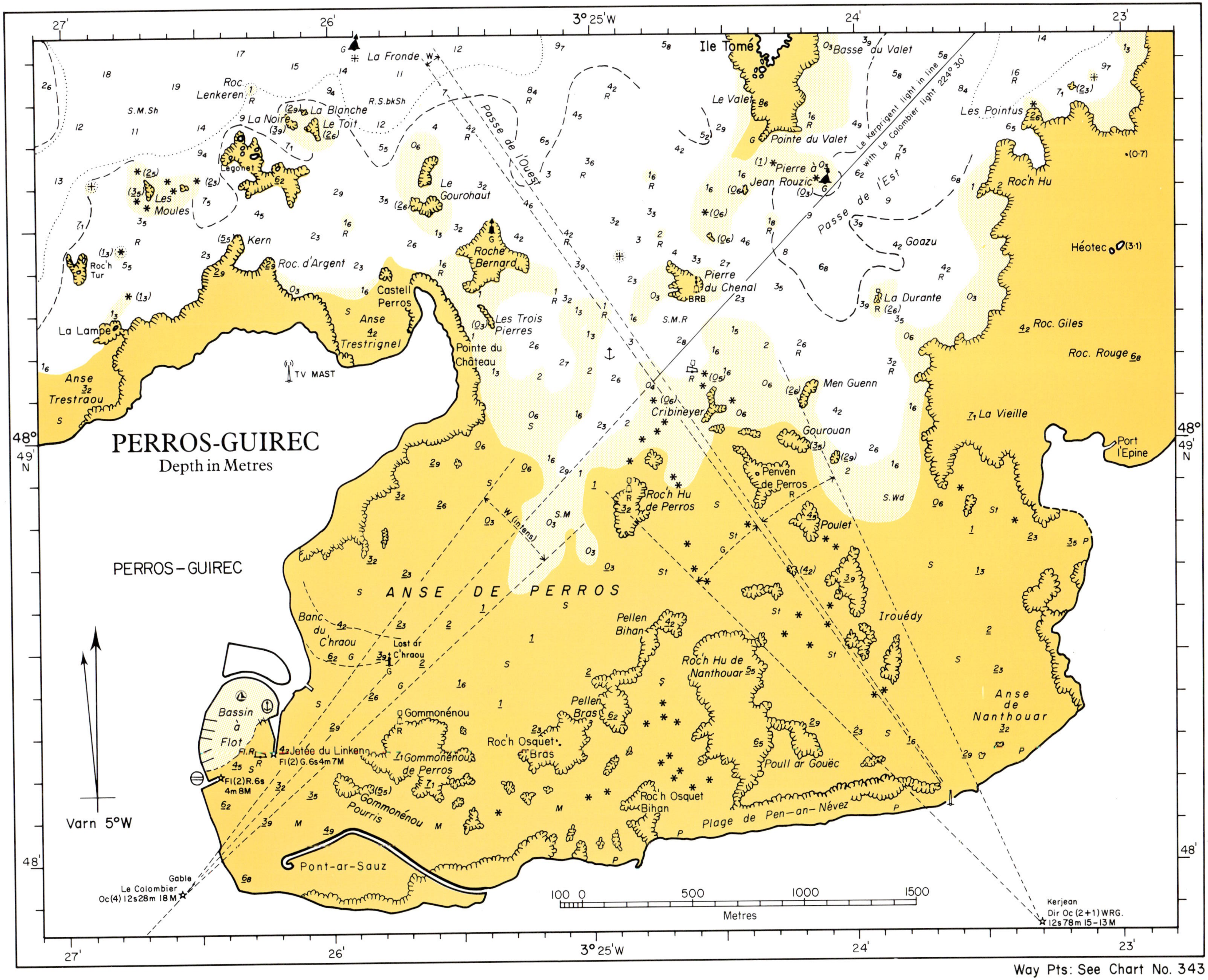

PERROS-GUIREC

Depth in Metres

PERROS-GUIREC

ANSE DE PERROS

Ile Tomé
Basse du Valet
Le Valet
Pointe du Valet
Pierre à Jean Rouzic
Passe de l'Est
Le Kerprigent light in line with Le Colombier light 224° 30'
Les Pointus
Roc'h Hu
Héotec
Goazu
La Durante
Roc. Giles
Roc. Rouge
La Vieille
Port l'Epine
Men Guenn
Gourouan
Pierre du Chenal
BRB
Cribineyer
Penven de Perros
Roc'h Hu de Perros
Poulet
Irouédy
Anse de Nanthouar
Pellen Bihan
Roc'h Hu de Nanthouar
Pellen Bras
Roc'h Osquet Bras
Roc'h Osquet Bihan
Poull ar Gouëc
Plage de Pen-an-Névez
La Fronde
Passe de l'Ouest
Roc. Lenkeren
La Blanche
La Noire
Le Toit
R.S.bkSh
S.M.Sh
Légonet
Les Moules
Le Gourohaut
Roche Bernard
Les Trois Pierres
Pointe du Château
Kern
Roc. d'Argent
Castell Perros
Anse Trestrignel
TV MAST
Roc'h Tur
La Lampe
Anse Trestraou
Banc du C'hraou
Lost ar C'hraou
Gommonénou
Gommonénou de Perros
Gommonénou Pourris
Bassin à Flot
Jetée du Linkenn
Fl(2)G.6s4m7M
Fl.R.
Fl(2)R.6s 4m8M
Pont-ar-Sauz
W (intens)
S.M.R
S.Wd
S.M

Gable
Le Colombier
Oc(4)12s28m 18M

Kerjean
Dir Oc(2+1)WRG.
12s78m 15-13M

Varn 5°W

100 0 500 1000 1500
Metres

27' 26' 3°25'W 24' 23'
48° 49' N
48'

Way Pts: See Chart No. 343

Ploumanac'h and Trégastel

See Chart No 343.

TIDE

Time and height differences on Standard Port (for instructions in use, see p. xii)
Note that this table, using times at St Helier as GMT, results in times at Ploumanac'h in French Standard Time (GMT +0100) without further adjustment.

	Lat	Long	Time Difference HW		LW		Height Difference (metres)			
ST HELIER See p. 114	49°11′N	2°07′W	0100 and 1300	0800 and 2000	0200 and 1400	0700 and 1900	11.1	8.1	4.1	1.3
Ploumanac'h	48°50′N	3°29′W	0000	+0005	−0025	0000	−2.1	−1.0	−0.7	−0.2

TIDES relative to HW Dover (approx) HW −0600. Stream just off Ploumanac'h sets E +0035; W +0610.

GENERAL

Ploumanac'h is a small drying harbour with a sill at 2.2m and berths for 180 including 20 visitors up to 12m LOA. The port is controlled from Perros-Guirec. Some facilities. Trégastel provides an anchorage in 3.5m, sheltered from all except onshore winds. Restaurants but no shops.

APPROACHES

Ploumanac'h
The entrance is marked by two unlit beacons and thence by other beacons.
Trégastel
Approach from the N between Le Taureau Beacon and Île Dhu. La Pierre Pendu is a distinctive hammer-head rock.

CAUTIONS

1 The approach should not be made at night or in strong onshore winds.
2 Do not anchor between the first beacons and the sill in Ploumanac'h.

TELEPHONES

Harbour Master: (96) 23 37 82

VHF

Perros-Guirec: Ch 16; 9

YACHT CLUB

Société Nautique de Perros-Guirec

The chart opposite is based on Chart No 3670 of the Service Hydrographique et Océanographique de la Marine (SHOM), reproduced by permission.

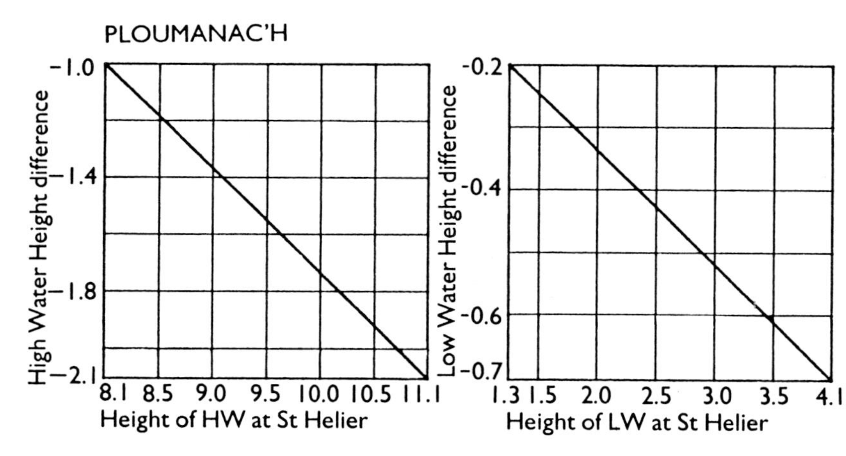

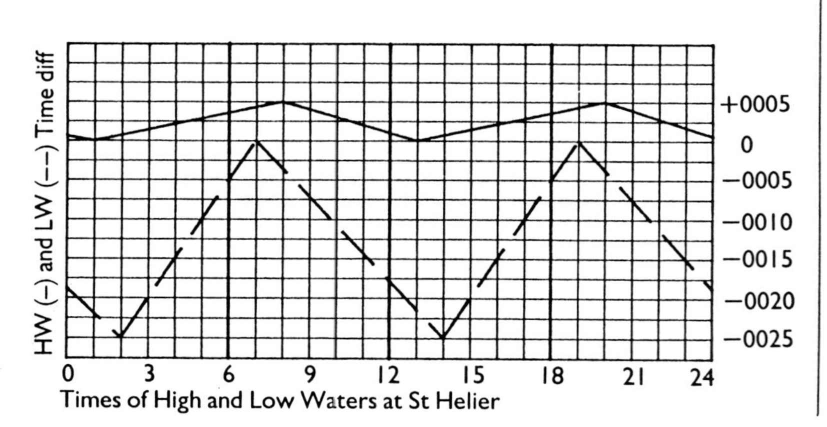

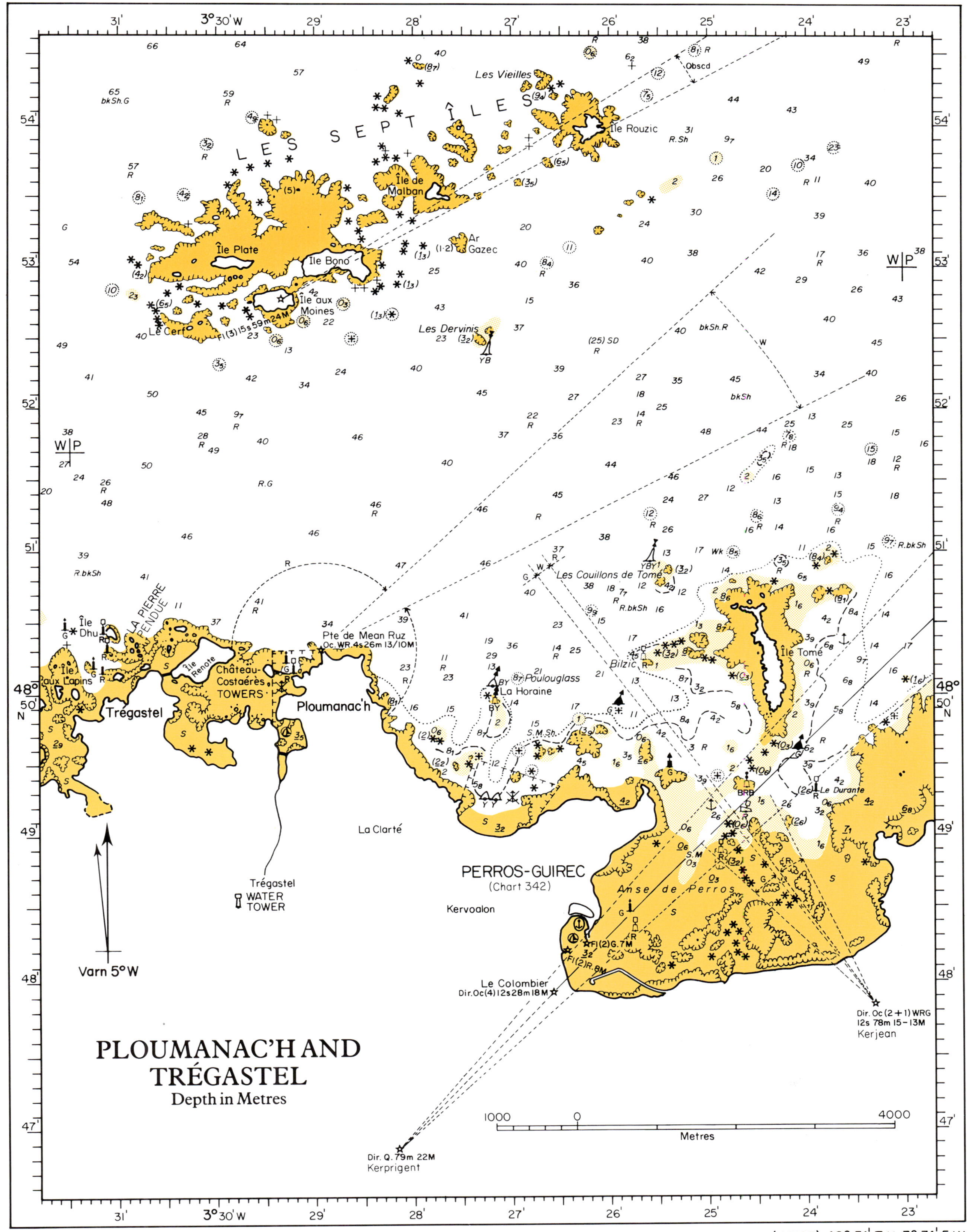
PLOUMANAC'H AND TRÉGASTEL
Depth in Metres
LES SEPT ÎLES
Les Vieilles
Île Rouzic
Île de Malban
Île Plate
Île Bono
Île aux Moines
Ar Gazec
Le Cerf
Fl(3)15s59m24M
Les Dervinis
YB
W|P
Île Dhu
Île aux Lapins
LA PIERRE PENDUE
Île Renote
Château-Costaérès TOWERS
Trégastel
Ploumanac'h
Pte de Mean Ruz
Oc.WR.4s26m 13/10M
Les Couillons de Tomé
YBY
Île Tomé
Bilzic
La Horaine
Poulouglass
Le Durante
La Clarté
PERROS-GUIREC
(Chart 342)
Kervoalon
Anse de Perros
Trégastel WATER TOWER
Varn 5°W
Fl(2)G.7M
Fl(2)R.8M
Le Colombier
Dir.Oc(4)12s28m18M
Dir. Oc (2+1) WRG
12s 78m 15-13M
Kerjean
Dir. Q. 79m 22M
Kerprigent
1000 0 4000
Metres
3°30'W 31' 29' 28' 27' 26' 25' 24' 23'
48°50'N 54' 53' 52' 51' 49' 48' 47'

Way Pts: (WEST) 48° 51'·7 N 3° 31'·5 W
(EAST) 48° 53'·0 N 3° 23'·0 W

Île Grande, Trébeurden and Lannion

See Chart No 344.

TIDE

Time and height differences on Standard Port (for instructions in use, see p. xii)

	Lat	Long	Time Difference HW		Time Difference LW		Height Difference (metres)			
BREST See p. 114	48°23′N	4°29′W	0000 and 1200	0600 and 1800	0000 and 1200	0600 and 1800	7.5	5.9	3.0	1.4
Trébeurden	48°46′N	3°35′W	+0105	+0110	+0120	+0100	+1.6	+1.3	+0.5	−0.1

TIDES relative to HW Dover (approx) HW +0620. Stream sets (½ mile W of Île Losquet) N +0115; S −0515.

GENERAL

The E side of the Baie de Lannion has a number of anchorages, sheltered in offshore winds. From the N, the most convenient are: (1) just N of Île Milliau, (2) just S of Pte de Bihit and (3) within the Rivière de Lannion. At neaps, a number of other anchorages can be found. There are shops, restaurants and hotels, principally at Trébeurden and Lannion, where there is also some provision for yachts.

APPROACHES

The approaches to (1) and (2) above are free of dangers and self-evident from the chart. There are some visitors' moorings at Trébeurden. The entrance to the Rivière de Lannion dries to 2.6m, rendering it dangerous in onshore winds. It is marked by unlit beacons. Once inside the river, there are a number of deep pools, created by dredging, which sounding will reveal; the most obvious one is off Le Yaudet. There is an Aero Radio Beacon in position 48°43′N 3°18′W 'LN' 345.5 kHz 50M.

CAUTION

1 There are many offlying rocks, some of which dry.

TELEPHONES

Harbour Master (Trébeurden):	(96) 23 66 93
Harbour Master (Lannion):	(96) 37 06 52
Customs:	(96) 37 45 32
Medical:	(96) 37 42 52 (Doctor, Lannion); (96) 23 50 66 (Doctor, Trébeurden)
British Consul:	(99) 46 26 64

YACHT CLUB

YC de Trébeurden: (96) 23 50 26

The chart opposite is based on Chart No 3669 of the Service Hydrographique et Océanographique de la Marine (SHOM), reproduced by permission.

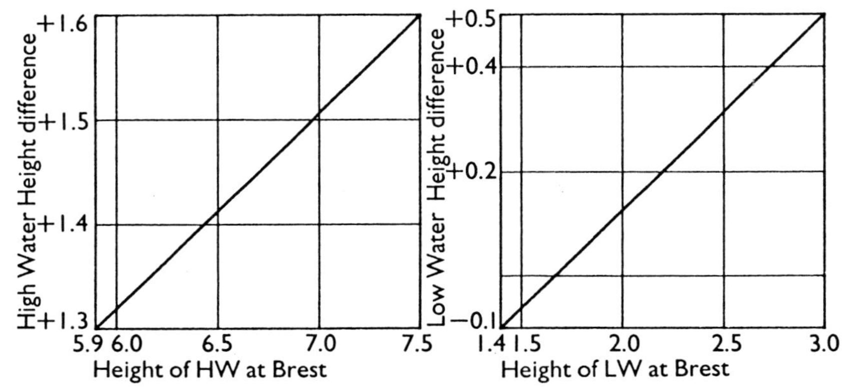

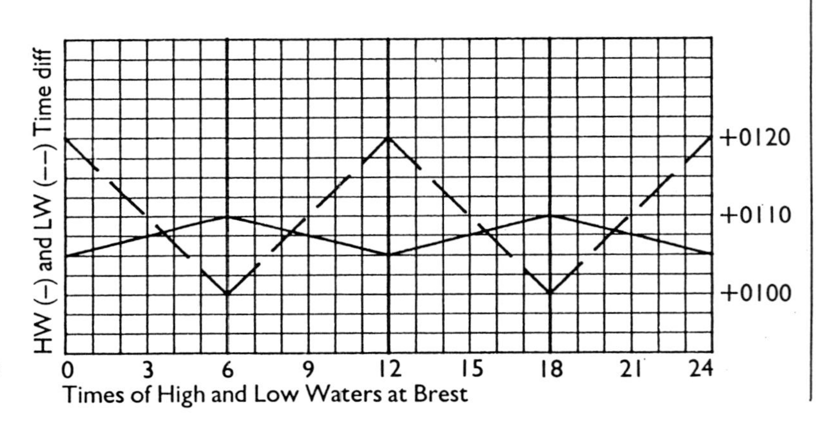

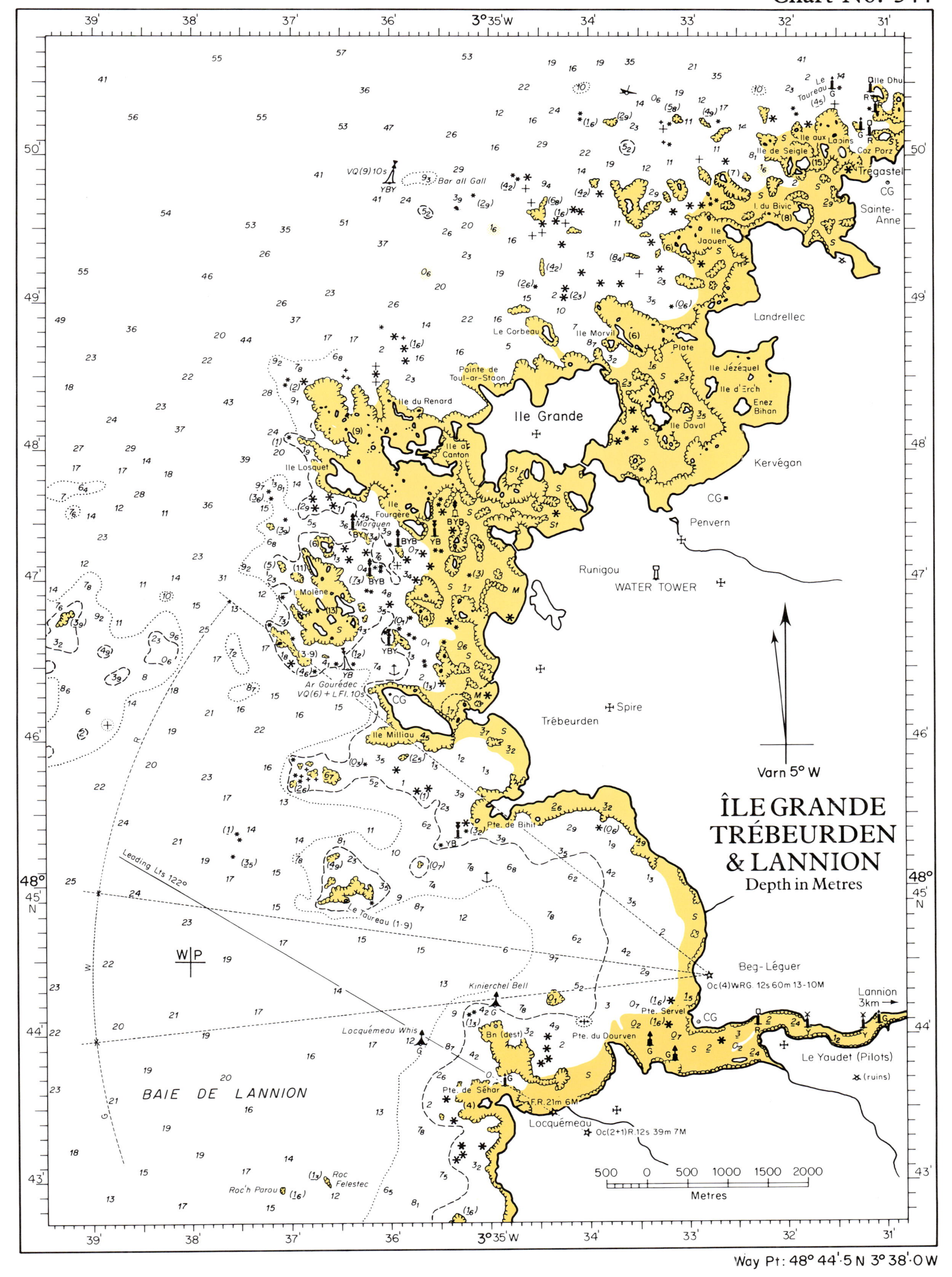
ÎLE GRANDE
TRÉBEURDEN
& LANNION
Depth in Metres
Varn 5° W
Ile Grande
Trébeurden
Trégastel
Sainte-Anne
Landrellec
Kervégan
Penvern
Runigou
WATER TOWER
Spire
Beg-Léguer
Oc(4)WRG. 12s 60m 13-10M
Lannion 3km
Le Yaudet (Pilots)
(ruins)
Locquémeau
Oc(2+1)R.12s 39m 7M
F.R.21m 6M
Pte. de Séhar
Pte. du Dourven
Pte. Servel
Kinierchel Bell
Locquémeau Whis
Bn (dest)
BAIE DE LANNION
Leading Lts 122°
Roc'h Parou
Roc Felestec
Le Taureau (1·9)
Pte. de Bihit
Ile Milliau
Ar Gourédec
VQ(6) + LFl. 10s
I. Molène
Ile Fougère
Morguen
Ile Losquet
Ile ar Canton
Ile du Renard
Pointe de Toul-ar-Staon
Le Corbeau
Ile Morvil
Plate
Ile Jézéquel
Ile d'Erc'h
Enez Bihan
Ile Daval
Ile Jaouen
I. du Bivic
Ile de Seigle
Ile aux Lapins
Coz Porz
Le Toureau
Ile Dhu
VQ(9) 10s
Bar all Gall
500 0 500 1000 1500 2000
Metres
Way Pt: 48° 44'·5 N 3° 38'·0 W

Primel and Baie de Morlaix

See Chart No 345.

TIDE

Time and height differences on Standard Port (for instructions in use, see p. xii)

	Lat	Long	Time Difference HW		Time Difference LW		Height Difference (metres)			
BREST See p. 114	48°23′N	4°29′W	0000 and 1200	0600 and 1800	0000 and 1200	0600 and 1800	7.5	5.9	3.0	1.4
Anse de Primel	48°43′N	3°50′W	+0100	+0110	+0115	+0055	+1.7	+1.3	+0.6	0.0
Château du Taureau	48°41′N	3°53′W	+0100	+0115	+0115	+0050	+1.5	+1.1	+0.5	−0.1

TIDES relative to HW Dover (approx) HW +0620. Stream sets (near Roches Jaunes) NE +0020; SW +0250.

GENERAL

Primel is a small harbour, well sheltered except in NW to N by E winds. Some facilities. The chart also shows the entrance to the Morlaix river.

APPROACH

Primel

The approach is clear of dangers and lit.

Rade de Morlaix

There are three channels: (1) Grand Chenal, passing E of Île Ricard and lit, (2) Grand Chenal, passing W of Île Ricard, with deeper water but unlit, and (3) Chenal de Tréguier, lit but partly drying. There is an anchorage in 6.3m off Penn Lan in the SW corner of the chart. See page 92 for passage up to Morlaix.

CAUTION

1 Primel may be dangerous in onshore winds.

TELEPHONES

Harbour Master: (98) 67 30 66

For others, see page 92.

YACHT CLUB

Centre Nautique: (98) 72 31 90

The chart opposite is based on Chart No 2745 of the Service Hydrographique et Océanographique de la Marine (SHOM), reproduced by permission.

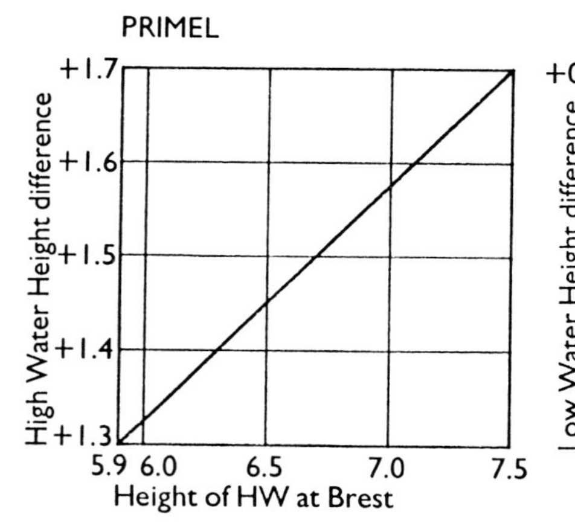

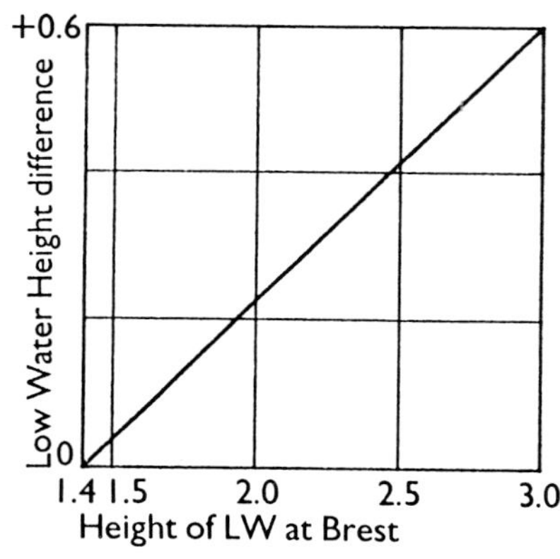

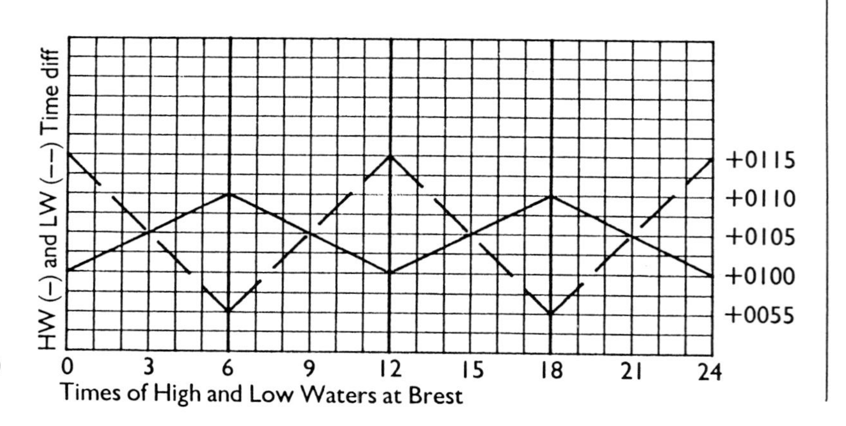

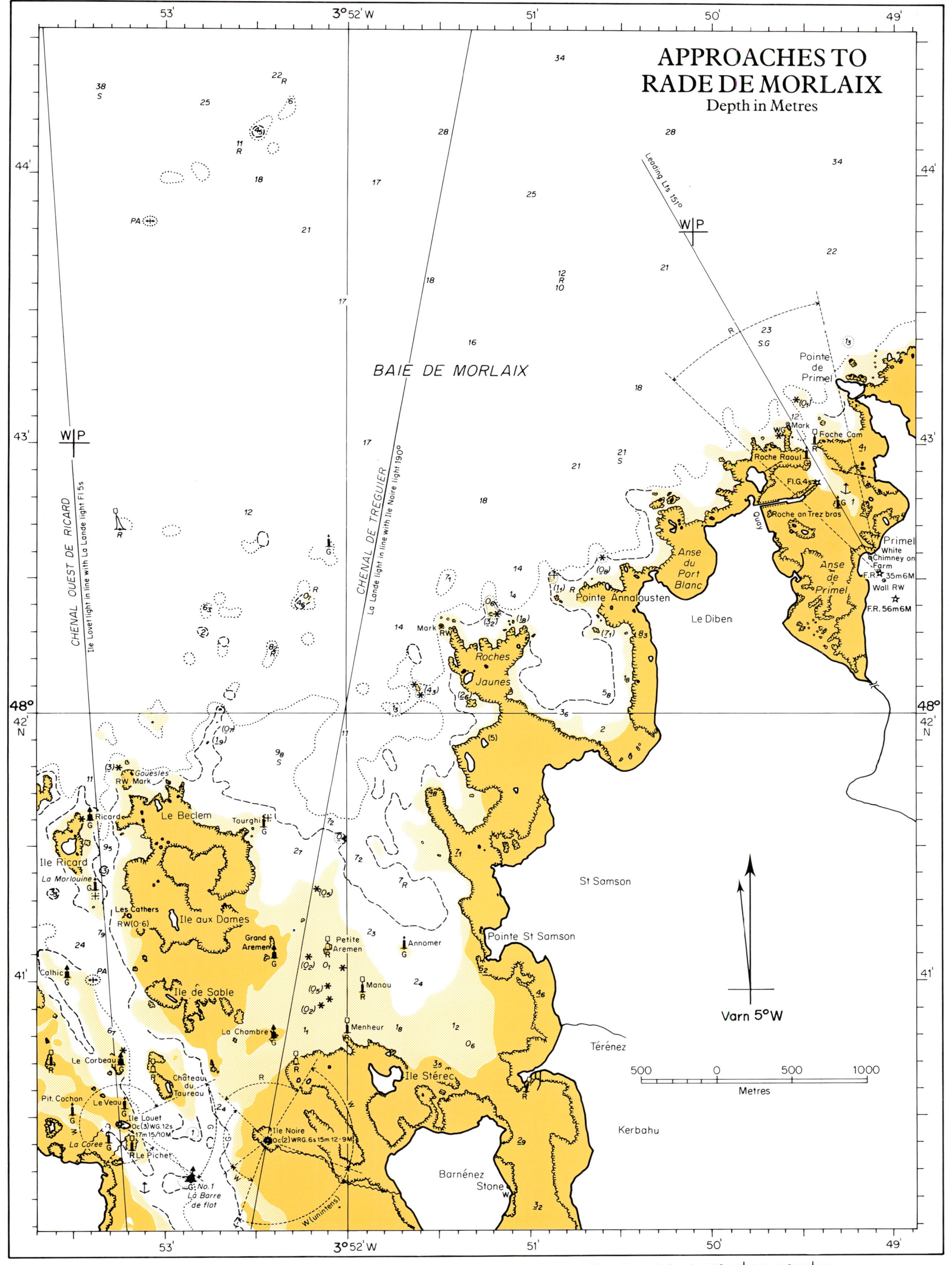

Way Pts: Primel 48° 43'·8N 3° 50'·1W

Chenal Ouest de Ricard ∠48° 43'·0N 3° 53'·5W

Morlaix

See Chart No 346.

TIDE

Time and height differences on Standard Port (for instructions in use, see p. xii)

	Lat	Long	Time Difference HW		Time Difference LW		Height Difference (metres)			
BREST See p. 114	48°23′N	4°29′W	0000 and 1200	0600 and 1800	0000 and 1200	0600 and 1800	7.5	5.9	3.0	1.4
Morlaix (Château du Taureau – see page 90)	48°41′N	3°53′W	+0100	+0115	+0115	+0050	+1.5	+1.1	+0.5	−0.1

TIDES relative to HW Dover (approx) See page 90.

GENERAL

The river is navigable by day to the locked basin at Morlaix, some 4½ miles from the entrance.

APPROACH

See page 90. The river is buoyed and then marked by beacons. On either side are oyster beds marked by withies. The lock opens 1½ hours before, at and 1 hour after HW during daylight, but in the season, probably more often.

TELEPHONES

Harbour Master:	(98) 62 13 14
Lock master:	(98) 88 54 92
Customs:	(98) 88 06 31
Medical:	(98) 88 40 22 (Hospital)
British Consul:	(99) 46 26 64

VHF

Port de Morlaix: Ch 9

MARINA

Morlaix: (98) 62 13 14

YACHT CLUB

YC de Morlaix: (98) 88 25 85

The chart opposite is based on Chart No 2745 of the Service Hydrographique et Océanographique de la Marine (SHOM), reproduced by permission.

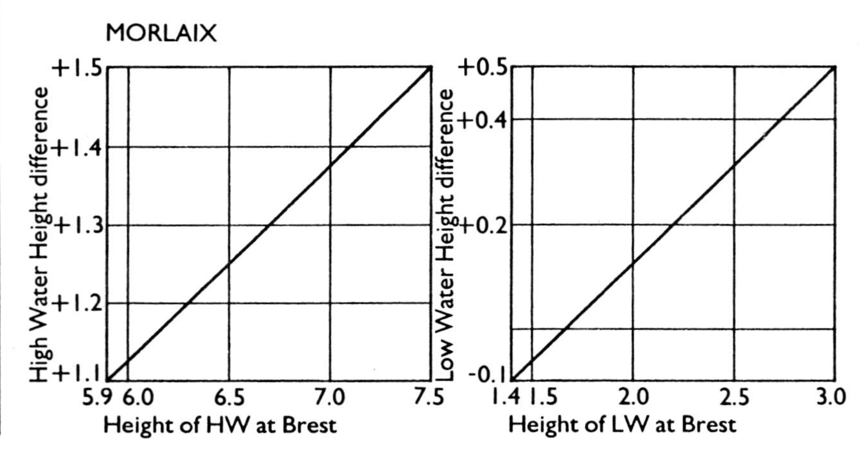

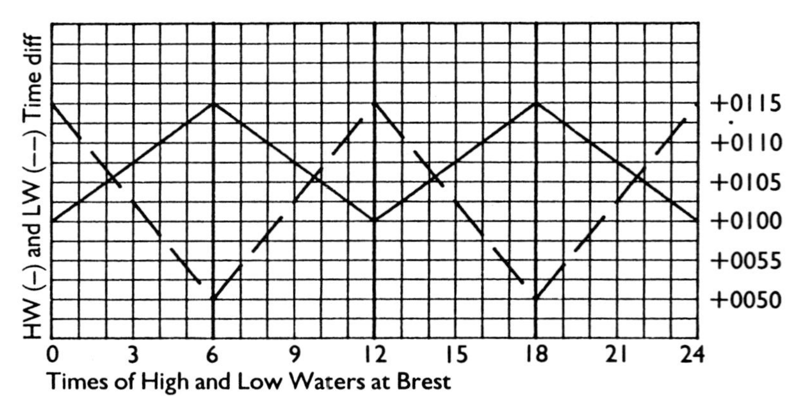

Chart No. 346

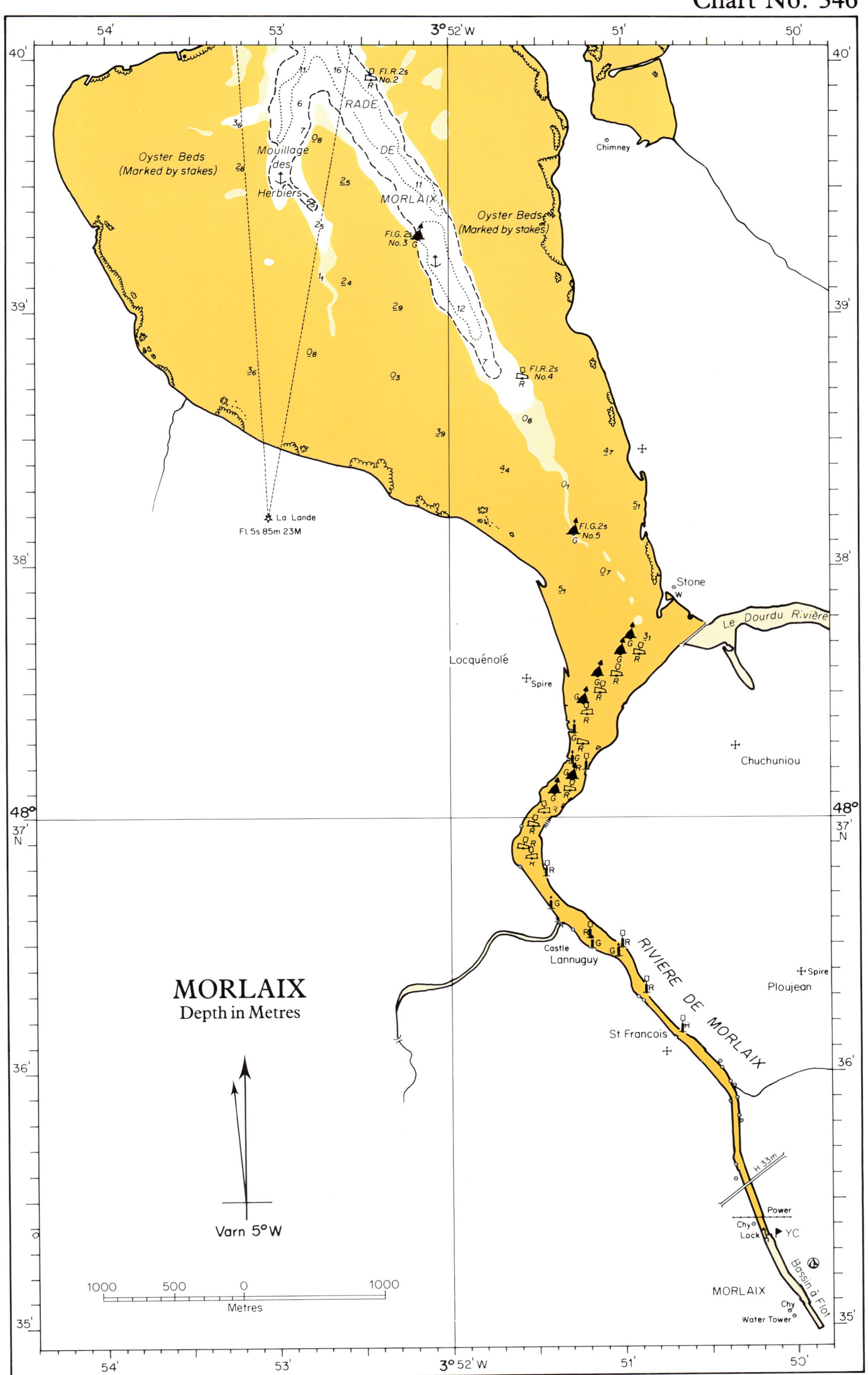
54'
53'
3° 52' W
51'
50'
40'
39'
38'
48°
37' N
36'
35'
Oyster Beds
(Marked by stakes)
Mouillage
des
Herbiers
RADE
DE
MORLAIX
Oyster Beds
(Marked by stakes)
Fl.R.2s
No.2
Fl.G.2s
No.3
Fl.R.2s
No.4
Fl.G.2s
No.5
Chimney
La Lande
Fl.5s 85m 23M
Stone
Le Dourdu R.vière
Locquénolé
Spire
Chuchuniou
Castle
Lannuguy
RIVIERE DE MORLAIX
Spire
Ploujean
St Francois
H 33m
Power
Chy
Lock
YC
Bassin à Flot
MORLAIX
Chy
Water Tower
MORLAIX
Depth in Metres
Varn 5°W
1000
500
0
1000
Metres

La Penzé Rivière

See Chart No 347.

TIDE

See page 96, Roscoff.

GENERAL

The river has two small towns accessible from yachts, St Pol-de-Leon with its harbour, Penpoul, on the W bank and Carantec on the E bank. The river and its approaches are unlit.

APPROACH

It is best to enter at about half flood when the many dangers are more apparent. The chart makes clear the possible approach tracks, but note that all of them have isolated rocks and none of them are fully marked. In clear, settled weather, however, sufficient leading lines are available to offer a safe passage. Penpoul has a dredged area in which it may be possible to anchor, the rest of the harbour dries. There is an anchorage in the river opposite the town but this is subject to swell in onshore winds. There is an anchorage in the river opposite Carantec. Limited yachting facilities in both towns.

CAUTIONS

1 There are many isolated dangers.
2 There are oyster beds on both sides of the river.
3 Secure shelter is only available well up the river.

The chart opposite is based on Chart No 2745 of the Service Hydrographique et Océanographique de la Marine (SHOM), reproduced by permission.

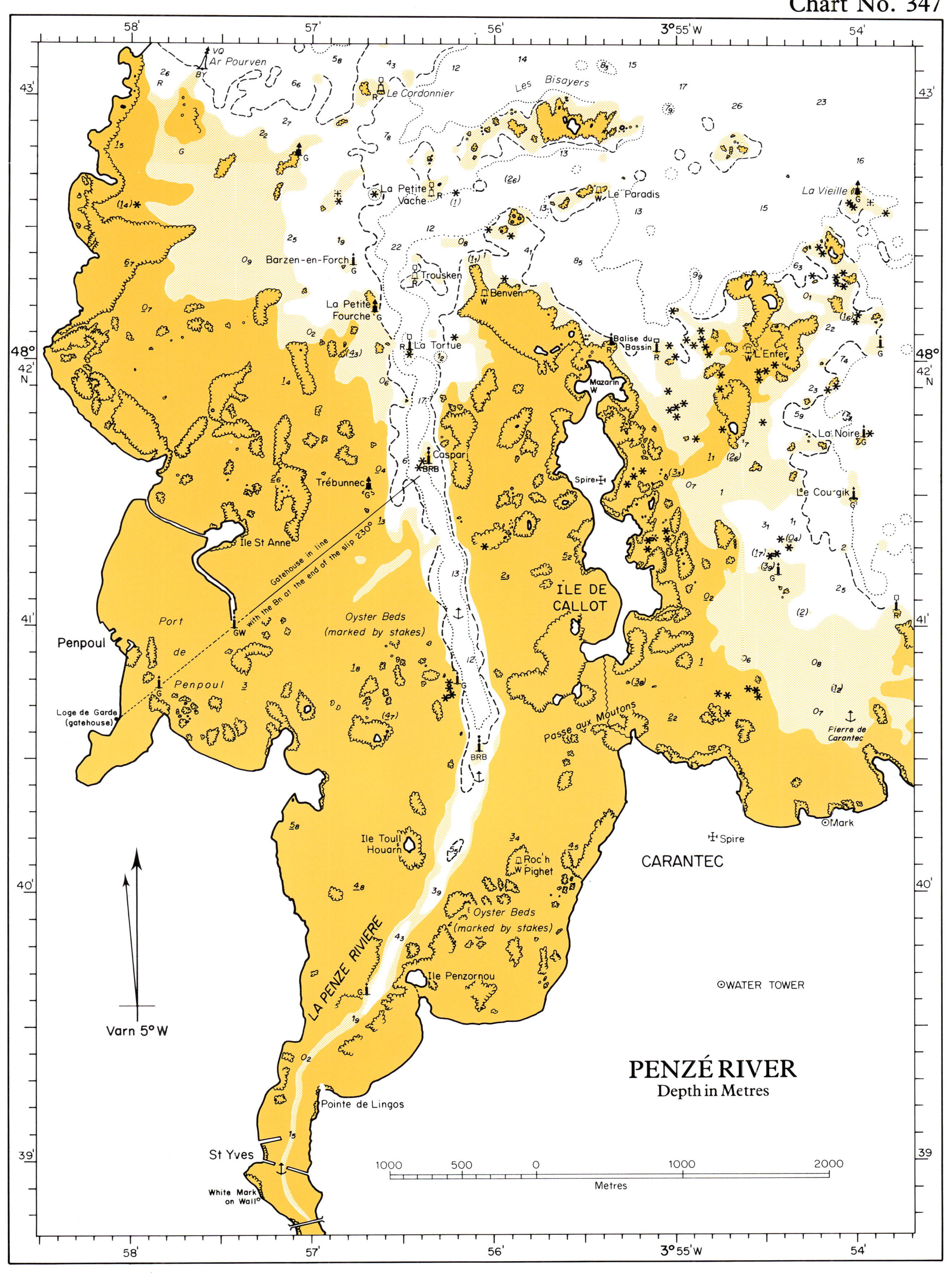

Ar Pourven
Le Cordonnier
Les Bisayers
La Petite Vache
Le Paradis
La Vieille
Barzen-en-Forch
Trousken
Benven
La Petite Fourche
La Tortue
Balise du Bassin
L'Enfer
Mazarin
La Noire
Caspari
Trébunnec
Spire
Le Cou-gik
Ile St Anne
Gatehouse in line with the Bn at the end of the slip 230°
ILE DE CALLOT
Port de Penpoul
Penpoul
Oyster Beds (marked by stakes)
Loge de Garde (gatehouse)
Passe aux Moutons
Fierre de Carantec
Mark
Spire
CARANTEC
Ile Toull Houarn
Roc'h Pighet
LA PENZE RIVIERE
Ile Penzornou
WATER TOWER
Varn 5° W
PENZÉ RIVER
Depth in Metres
Pointe de Lingos
St Yves
White Mark on Wall
1000 500 0 1000 2000
Metres

Roscoff and Île de Batz

See Chart No 348.

TIDE

Time and height differences on Standard Port (for instructions in use, see p. xii)

	Lat	Long	Time Difference HW		LW		Height Difference (metres)			
BREST See p. 114	48°23′N	4°29′W	0000 and 1200	0600 and 1800	0000 and 1200	0600 and 1800	7.5	5.9	3.0	1.4
Roscoff	48°43′N	3°58′W	+0055	+0105	+0115	+0050	+1.4	+1.1	+0.5	−0.1

TIDES relative to HW Dover (approx) HW −0610. Stream sets (in the channel between Île de Batz and the mainland) E +0035; W −0605.

GENERAL

Roscoff is a drying harbour, somewhat difficult of access but with good shelter except in strong NE winds; all facilities. Bloscon is a ferry port and closed to yachts (except by special permission from the Harbour Master), though a deep water anchorage is available S of the port limits. On the Île de Batz, there is a small drying harbour, Porz Kernoch, with limited facilities and a ferry service to Roscoff.

APPROACH

Bloscon is easy of approach, but yachts must not anchor N of the N cardinal buoy marking the SE port limit. There is a Radio Beacon (BC 287.3 kHz 10M). Roscoff should be approached with great care down the 210° leading line and not earlier than −3HW, noting the isolated rocks; berth in the Vieux Port, dries 4.9m. Anchor in the channel as shown, though see below.

CAUTIONS

1 Keep clear of the port limits of Bloscon.
2 The stream runs very strongly in the channel, nearly four knots, and 'wind against tide' produces an uncomfortable berth.
3 The whole area is encumbered with isolated rocks and rocky ledges. Great care is needed.

TELEPHONES

Harbour Master (Roscoff):	(98) 69 19 59
(Bloscon):	(98) 61 27 84
Customs:	(98) 61 27 86
Medical:	(98) 69 71 18 (Doctor)
British Consul:	(99) 46 26 64

VHF

Roscoff and Bloscon: Ch 12; 16

YACHT CLUB

Cercle Nautique de Roscoff: (98) 69 72 79

The chart opposite is based on Chart No 2745 of the Service Hydrographique et Océanographique de la Marine (SHOM), reproduced by permission.

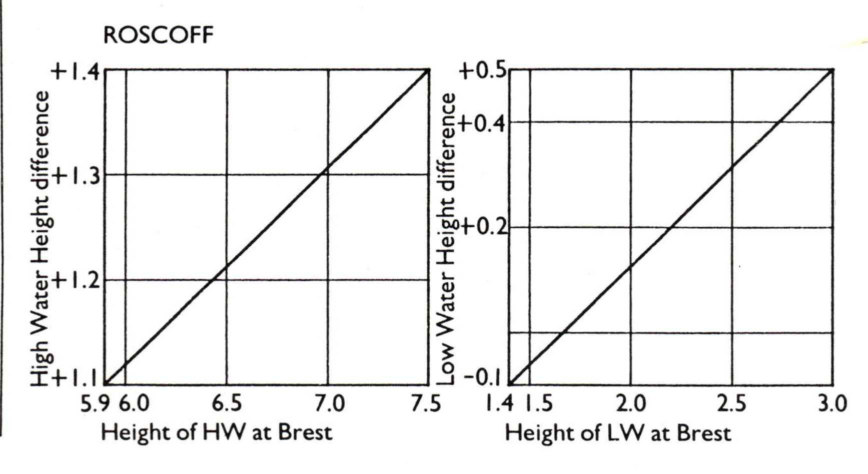

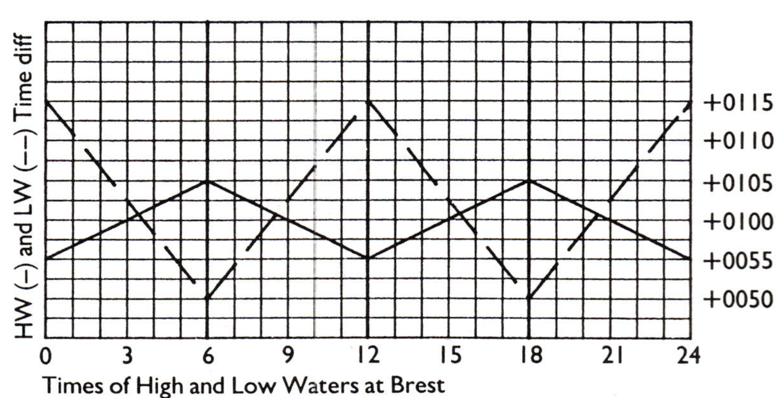

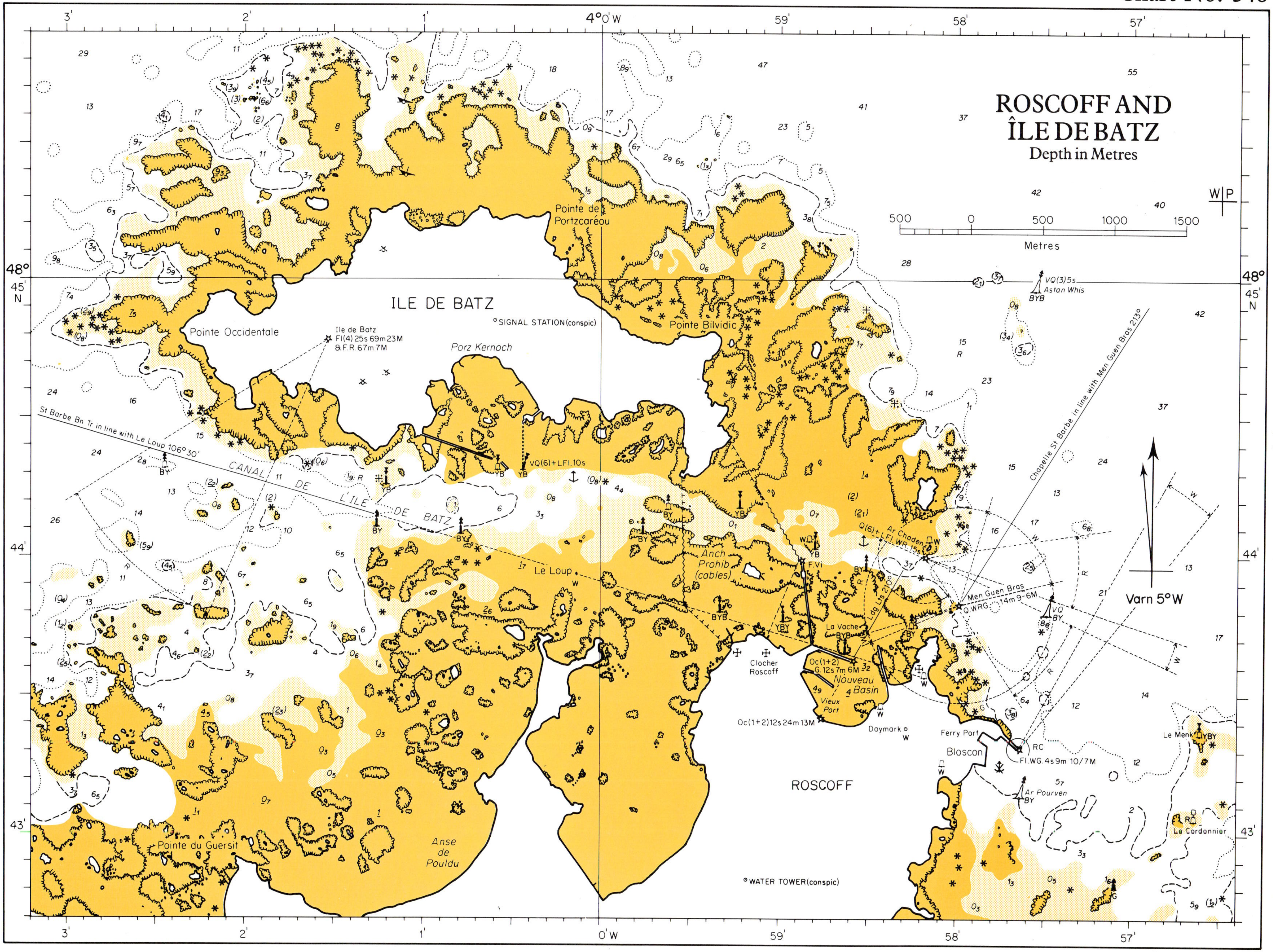

ROSCOFF AND ÎLE DE BATZ

Depth in Metres

Metres 500 0 500 1000 1500

ILE DE BATZ

Pointe Occidentale

Ile de Batz Fl(4) 25s 69m 23M & F.R. 67m 7M

°SIGNAL STATION (conspic)

Porz Kernoch

Pointe de Portzcaréou

Pointe Bilvidic

VQ(3)5s Astan Whis BYB

Chapelle St Barbe in line with Men Guen Bras 213°

St Barbe Bn Tr in line with Le Loup 106°30'

CANAL DE L'ILE DE BATZ

VQ(6)+L Fl. 10s

Le Loup

Anch Prohib (cables)

Q(6)+L Fl. WR. 15s Ar Chaden

Men Guen Bras Q.WRG. 14m 9-6M

VQ BY

Ldg Lts 210°

La Vache BYB

F.Vi

Oc(1+2) G.12s 7m 6M

Nouveau Basin

Vieux Port

Clocher Roscoff

Oc(1+2) 12s 24m 13M

Daymark W

Ferry Port

Bloscon

RC Fl.WG. 4s 9m 10/7M

Ar Pourven BY

Le Menk YBY

RC Le Cardonnier

ROSCOFF

°WATER TOWER (conspic)

Anse de Pouldu

Pointe du Guersit

Varn 5°W

Way Pts: (East) 48° 45'·3N 3°56'·5W
(West) (5 cables outside border on 106° 30'
Ldg Line) 48°44'·7N 4°04'·0W

Pontusval

See Chart No 349.

TIDE

See page 100.
TIDES relative to HW Dover (approx) HW +0605. Stream sets E +0110; W −0515.

GENERAL

Pontusval is a drying harbour only really usable by bilge keeled yachts or yachts with legs and then only in settled weather. Its pleasant character may be sufficient to compensate for its undeniable drawbacks. There are shops at Brignogan.

APPROACH

178° on Coat Tanguy beacon (in line with spire 1¼ miles S) leads clear (but only just). There are three rocks painted white to be left to starboard.

CAUTION

1 The harbour is dangerous in onshore winds.

The chart opposite is based on Chart No 3668 of the Service Hydrographique et Océanographique de la Marine (SHOM), reproduced by permission.

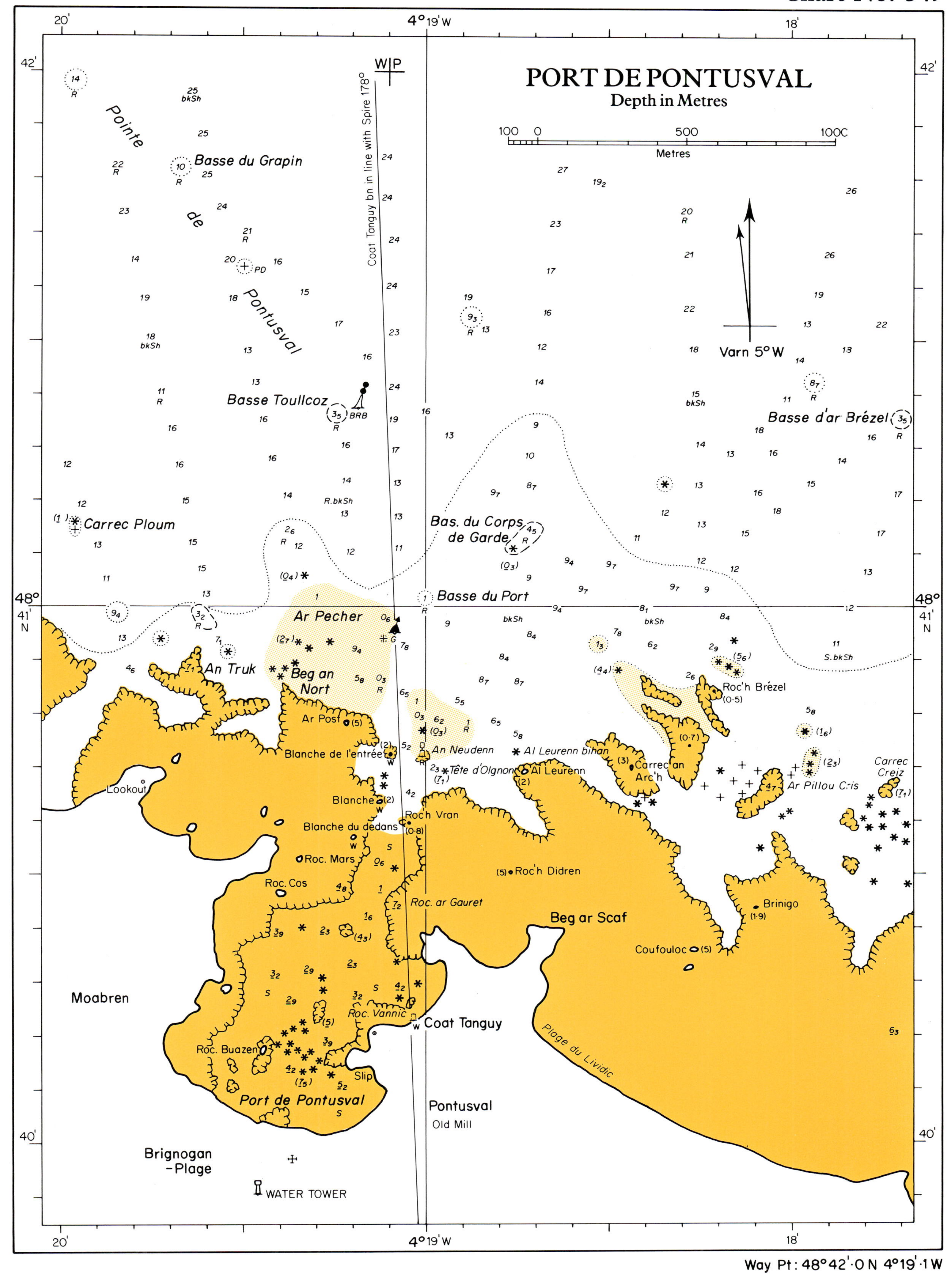
PORT DE PONTUSVAL
Depth in Metres
Metres
Varn 5°W
W/P
Coat Tanguy bn in line with Spire 178°
Pointe de Pontusval
Basse du Grapin
Basse Toullcoz
Carrec Ploum
Bas. du Corps de Garde
Basse du Port
Basse d'ar Brézel
Ar Pecher
An Truk
Beg an Nort
Ar Post
Blanche de l'entrée
Blanche
Blanche du dedans
An Neudenn
Tête d'Olgnon
Al Leurenn bihan
Al Leurenn
Roc'h Vran
Roc. Mars
Roc. Cos
Roc. ar Gauret
Roc'h Didren
Roc'h Brézel
Carrec an Arc'h
Ar Pillou Cris
Carrec Creiz
Beg ar Scaf
Brinigo
Coufouloc
Plage du Lividic
Lookout
Moabren
Roc. Vannic
Coat Tanguy
Roc. Buazen
Slip
Port de Pontusval
Pontusval
Old Mill
Brignogan -Plage
WATER TOWER
Way Pt: 48°42'·0 N 4°19'·1 W

L'Aber Vrac'h

See Chart No 350.

TIDE

Time and height differences on Standard Port (for instructions in use, see p. xii)

	Lat	Long	Time Difference HW		Time Difference LW		Height Difference (metres)			
BREST See p. 114	48°23′N	4°29′W	0000 and 1200	0600 and 1800	0000 and 1200	0600 and 1800	7.5	5.9	3.0	1.4
Île Ćezon	48°36′N	4°35′W	+0020	+0030	+0035	+0020	+0.5	+0.2	−0.1	−0.3

TIDES relative to HW Dover (approx) HW +0540. Stream sets NE +0010; WSW +0610.

GENERAL

L'Aber Vrac'h provides good shelter, pontoon berths and an anchorage. All facilities for yachtsmen are available.

APPROACH

Use the Grand Chenal which is well marked and lit, but the leading lines should be carefully observed as there are unlit dangers on either side. Note that Île Vierge has a Radio Beacon (VG 298.8 kHz 70M). Use the facilities at L'Aber or go about a mile further up the river and anchor in sheltered deep water. It is possible to go further up to Paluden.

CAUTIONS

1 There are oyster beds to be avoided.
2 Note the dangers on either side of the approach line.

TELEPHONES

Harbour Master:	(98) 04 91 62
Customs:	(98) 04 90 27
Medical:	(98) 04 91 87 (Doctor)
British Consul:	(99) 46 26 64

VHF

L'Aber Vrac'h: Ch 9 (0700 to 2100 local time)

MARINA

See Harbour Master

YACHT CLUB

YC des Abers: (98) 04 92 60

The chart opposite is based on Chart No 1432 of the Service Hydrographique et Océanographique de la Marine (SHOM), reproduced by permission.

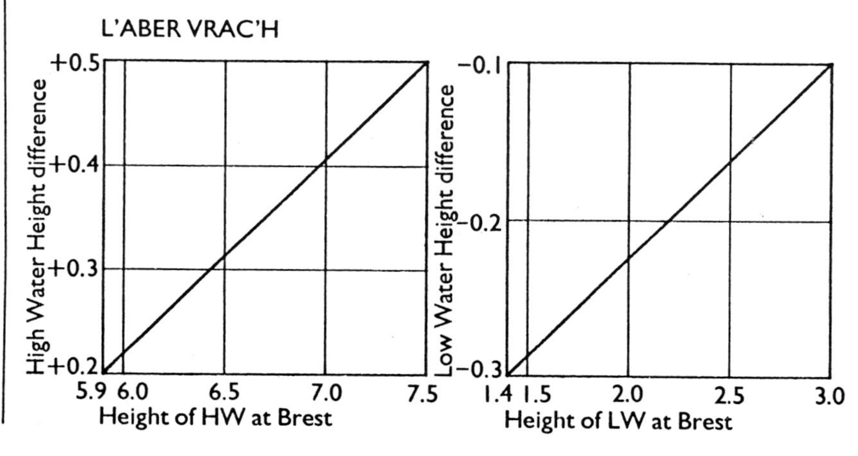

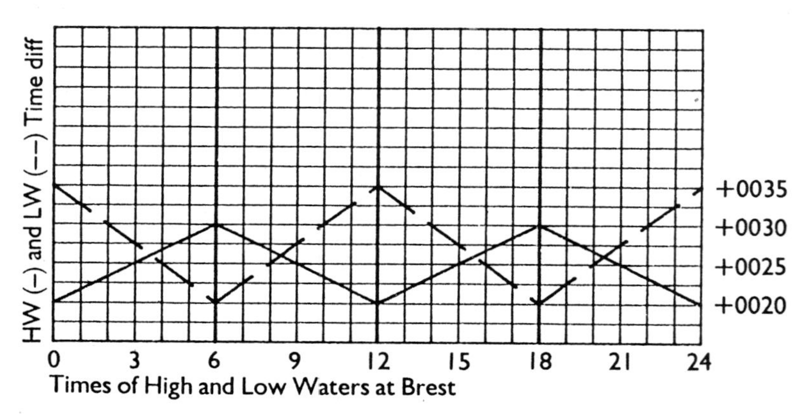

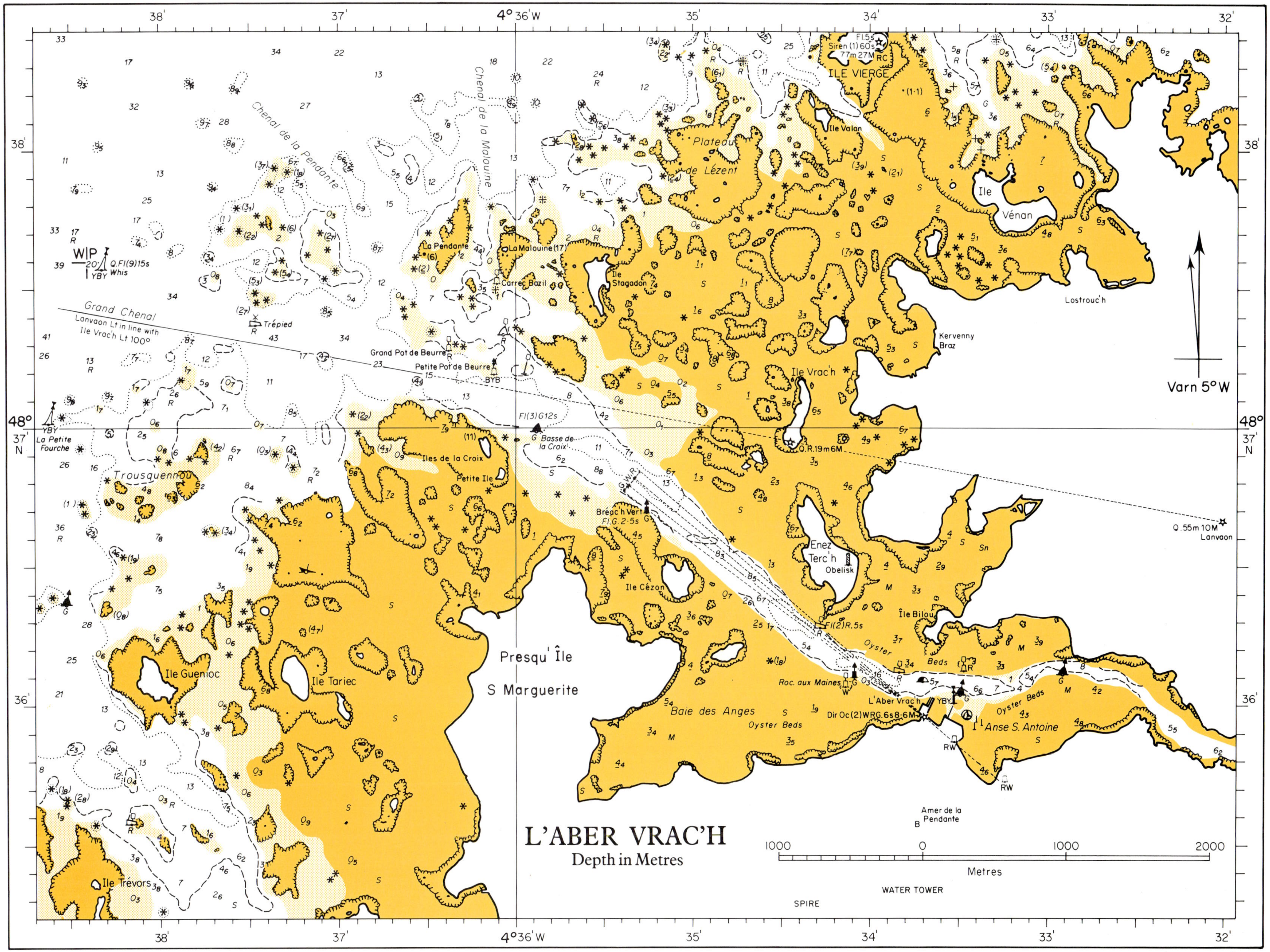

L'ABER VRAC'H

Depth in Metres

Way Pt: 48°37'·6 N 4°38'·4 W

Varn 5°W

1000 0 1000 2000
Metres

ILE VIERGE
Fl.5s
Siren (1) 60s
77m 27M RC
Ile Valan
Plateau de Lézent
Ile Vénan
Lostrouc'h
Kervenny Braz
Ile Vrac'h
Q.R.19m6M
Ile Stagadon
La Malouine (17)
Carrec Bazil
La Pendante
Chenal de la Malouine
Chenal de la Pendante
Grand Chenal
Lanvaon Lt in line with Ile Vrac'h Lt 100°
Q.55m 10M
Lanvaon
W/P
Q.Fl(9)15s
Whis
YBY
Trépied
Grand Pot de Beurre
Petite Pot de Beurre
BYB
Fl(3)G12s
G Basse de la Croix
La Petite Fourche
Trousquennou
Iles de la Croix
Petite Ile
Bréac'h Vert
Fl.G.2·5s
Ile Cézon
Enez Terc'h
Obelisk
Fl(2)R.5s
Ile Bilou
Oyster Beds
Roc. aux Maines
L'Aber Vrac'h
Dir Oc(2)WRG.6s8·6M
Anse S. Antoine
Baie des Anges
Presqu' Île S Marguerite
Ile Tariec
Ile Guenioc
Ile Trévors
Amer de la Pendante
WATER TOWER
SPIRE

L'Aber Benoit

See Chart No 351.

TIDE

Time and height differences on Standard Port (for instructions in use, see p. xii)

	Lat	Long	Time Difference HW		Time Difference LW		Height Difference (metres)			
			0000 and 1200	0600 and 1800	0000 and 1200	0600 and 1800				
BREST See p. 114	48°23′N	4°29′W					7.5	5.9	3.0	1.4
L'Aber Benoit	48°35′N	4°38′W	+0020	+0020	+0035	+0035	+0.6	+0.5	+0.1	−0.2

TIDES relative to HW Dover (approx) HW +0545.
Stream sets (off the entrance) ENE −0005; W +0605 (in the river) Floods +0025; Ebbs +0610.

GENERAL

The river offers good and quiet shelter but is only accessible in daylight, good visibility and absence of strong onshore winds. No special yachting facilities.

APPROACH

Enter either side of the Plateau Ruzven as both are marked. Thence by buoys and beacons. None of the marks are lit.

CAUTIONS

1 The entrance may be dangerous in strong onshore winds.
2 There are oyster beds.

TELEPHONES

See page 100.

The chart opposite is based on Chart No 1432 of the Service Hydrographique et Océanographique de la Marine (SHOM), reproduced by permission.

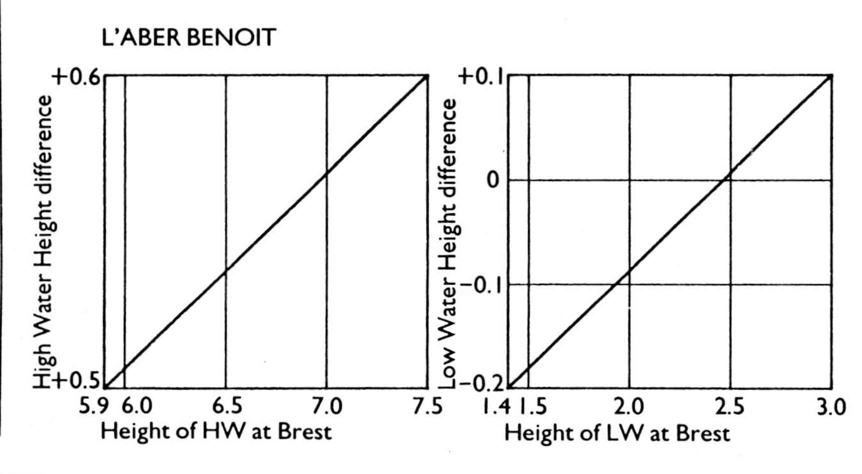

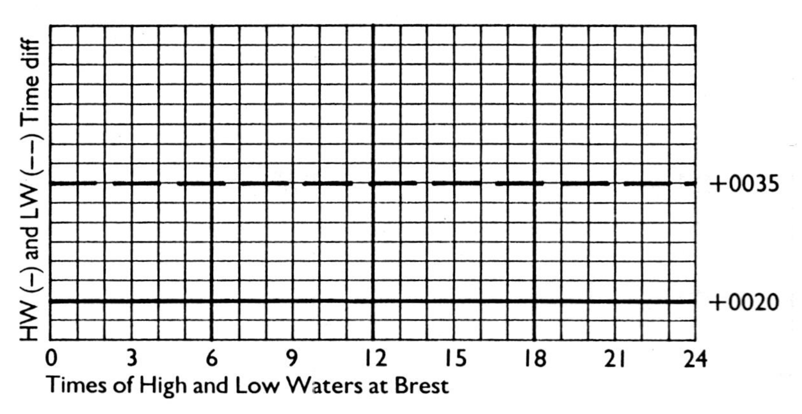

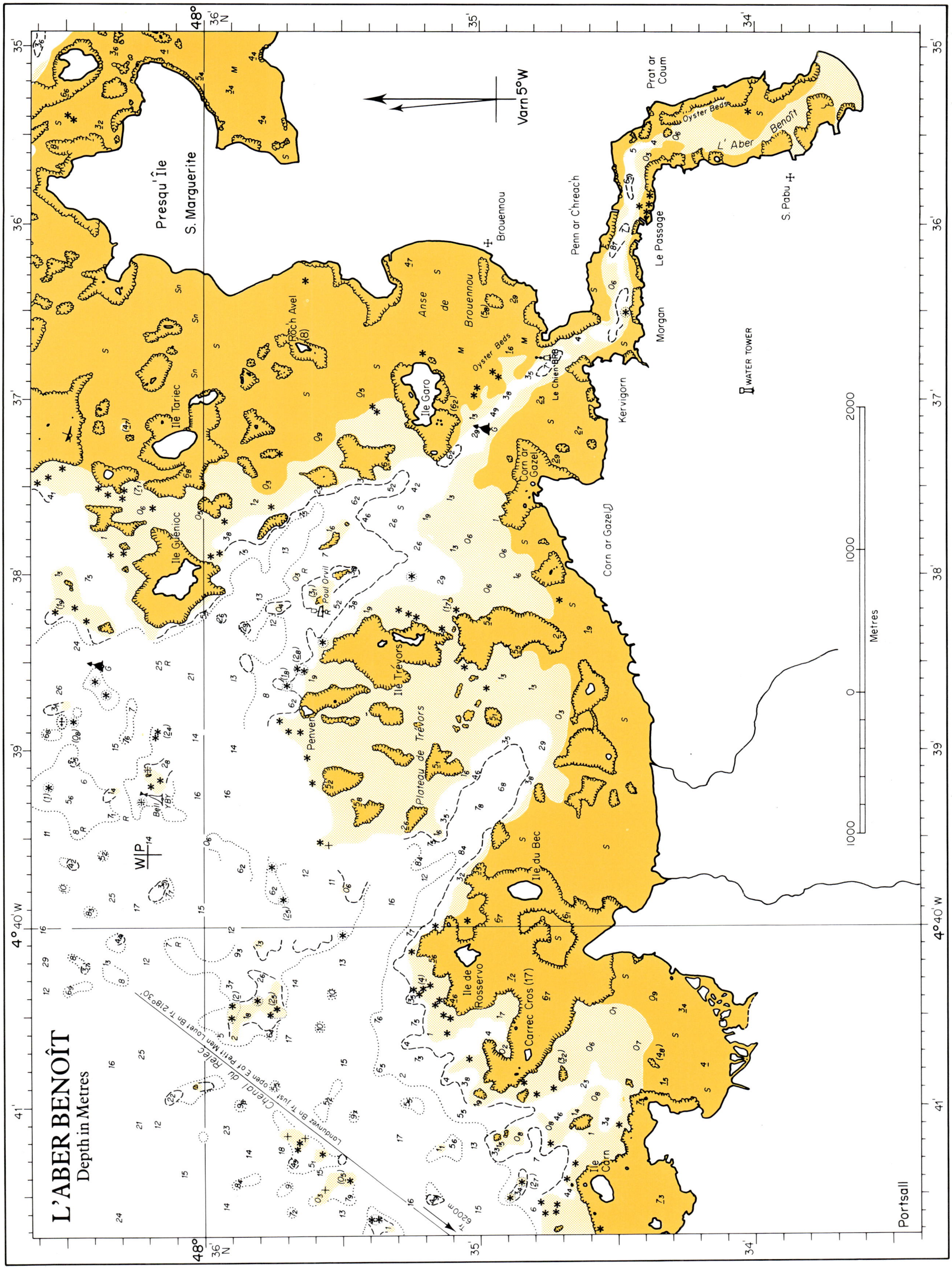

L'ABER BENOÎT

Depth in Metres

Portsall

See Chart No 352.

TIDE

Time and height differences on Standard Port (for instructions in use, see p. xii)

	Lat	Long	Time Difference HW		LW		Height Difference (metres)			
			0000 and 1200	0600 and 1800	0000 and 1200	0600 and 1800				
BREST See p. 114	48°23′N	4°29′W					7.5	5.9	3.0	1.4
Portsall	48°33′N	4°42′W	+0020	+0020	+0020	+0010	+0.6	+0.6	+0.4	+0.1

TIDES relative to HW Dover (approx) HW +0525.
Stream sets (at Grande Basse de Portsall lt buoy) NE +0040; WSW −0450.
(at Basse Paupian buoy) E +0005; SW +0610.

GENERAL

Portsall is a small drying harbour. Anchor off or lie alongside the jetty if there is room. Limited shops but Ploudalmezeau (2 miles by bus) has good shops and restaurants.

APPROACH

Down the W sector (084°–088°) of Portsall Lt, then between the two beacon towers.

CAUTIONS

1 The harbour is exposed to onshore winds.
2 Note the Entry Prohibited area round the wreck of the tanker, *Amoco Cadiz*. Diving is prohibited within the Restricted area.

TELEPHONES

Harbour Master:	(98) 80 62 25
Medical:	(98) 48 10 46

YACHT CLUB

CN de Portsall-Kersaint:	(98) 48 63 10

The chart opposite is based on Chart No 1432 of the Service Hydrographique et Océanographique de la Marine (SHOM), reproduced by permission.

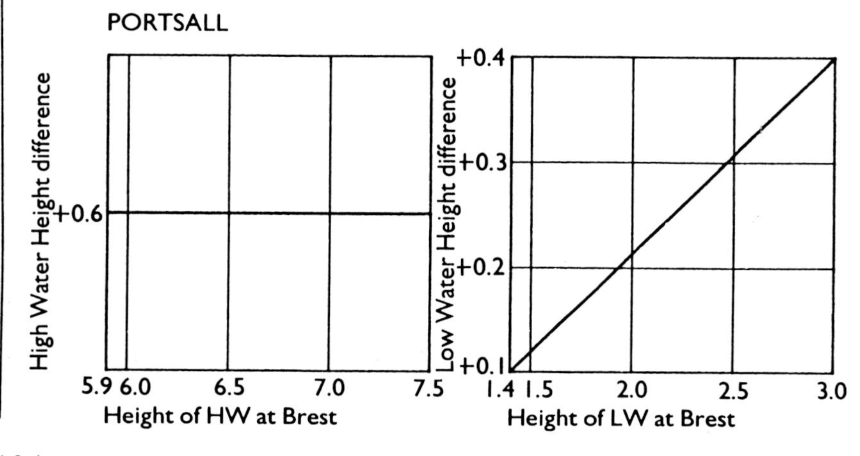

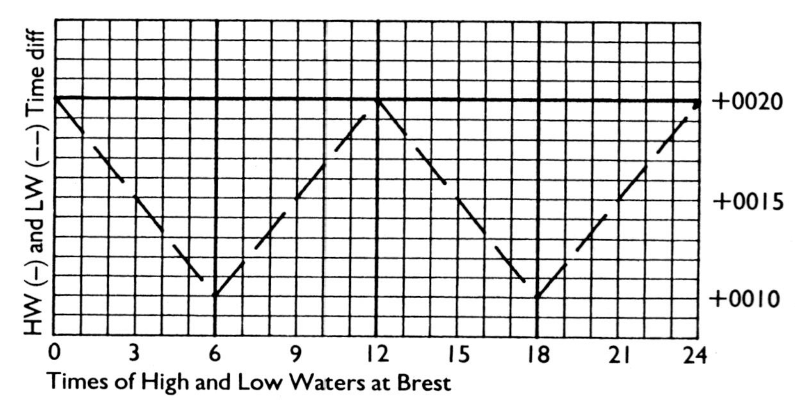

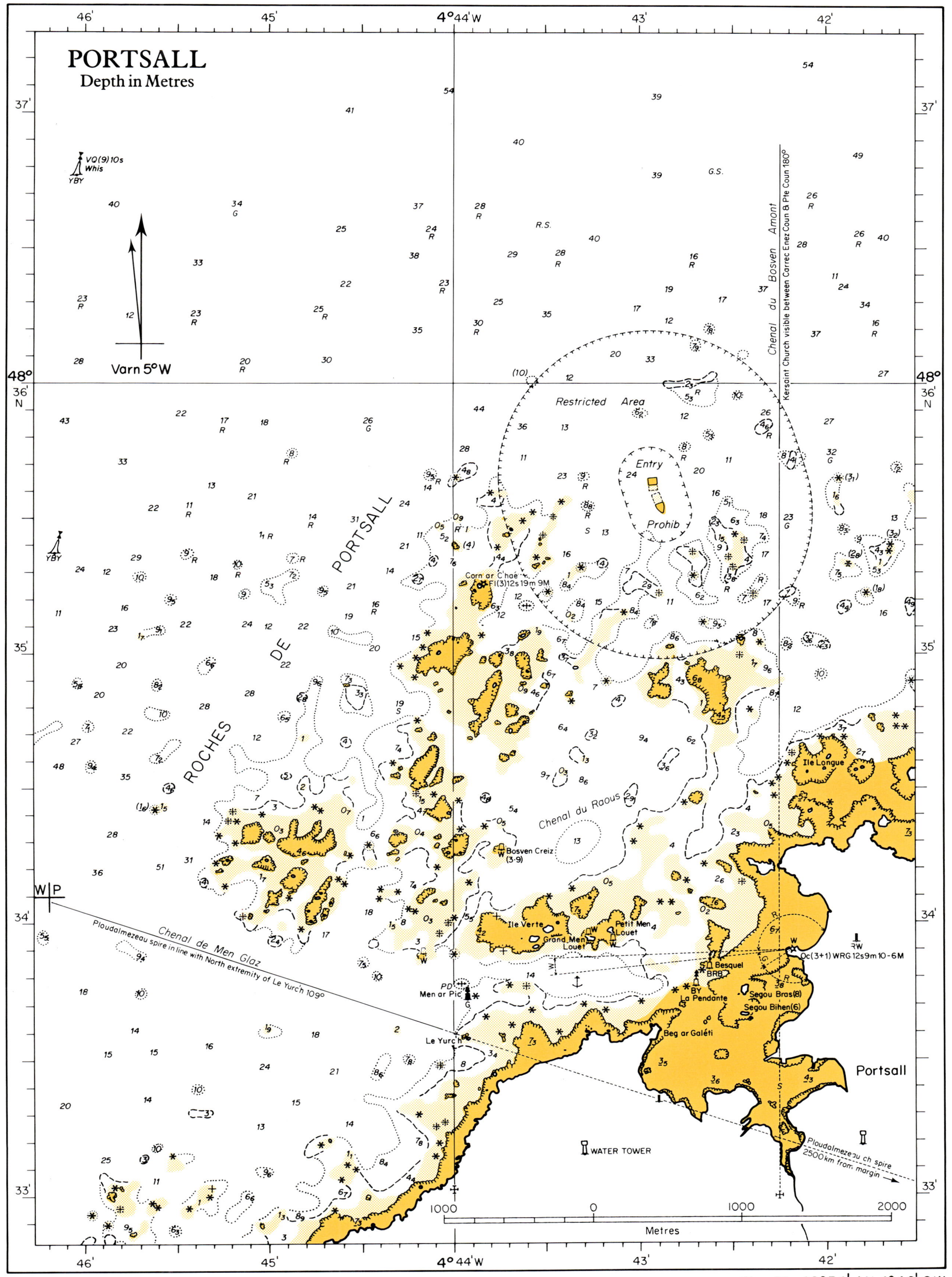

Way Pt: 48°34'·1N 4°46'·2W

L'Aberildut

See Chart No 353.

TIDE

See page 108.

GENERAL

L'Aberildut is a small harbour, originally drying but now with a deep pool created by dredging. There is a hard drying bar. Some facilities.

APPROACH

Use the leading line shown and note Le Lieu beacon tower. At night there is a W sectored light. A port hand and a starboard hand beacon mark roughly the extremities of the deep dredged channel. Anchor off between SE and E of Le Lieu or, within the harbour, preferably SSW about 100m from Laber slip, or wherever there is room. It is advisable to buoy the anchor.

CAUTIONS

1 The approach needs care as there are rocks on either side of the entrance.
2 The stream runs strongly across the entrance.

The chart opposite is based on Chart No 3345 of the Service Hydrographique et Océanographique de la Marine (SHOM), reproduced by permission.

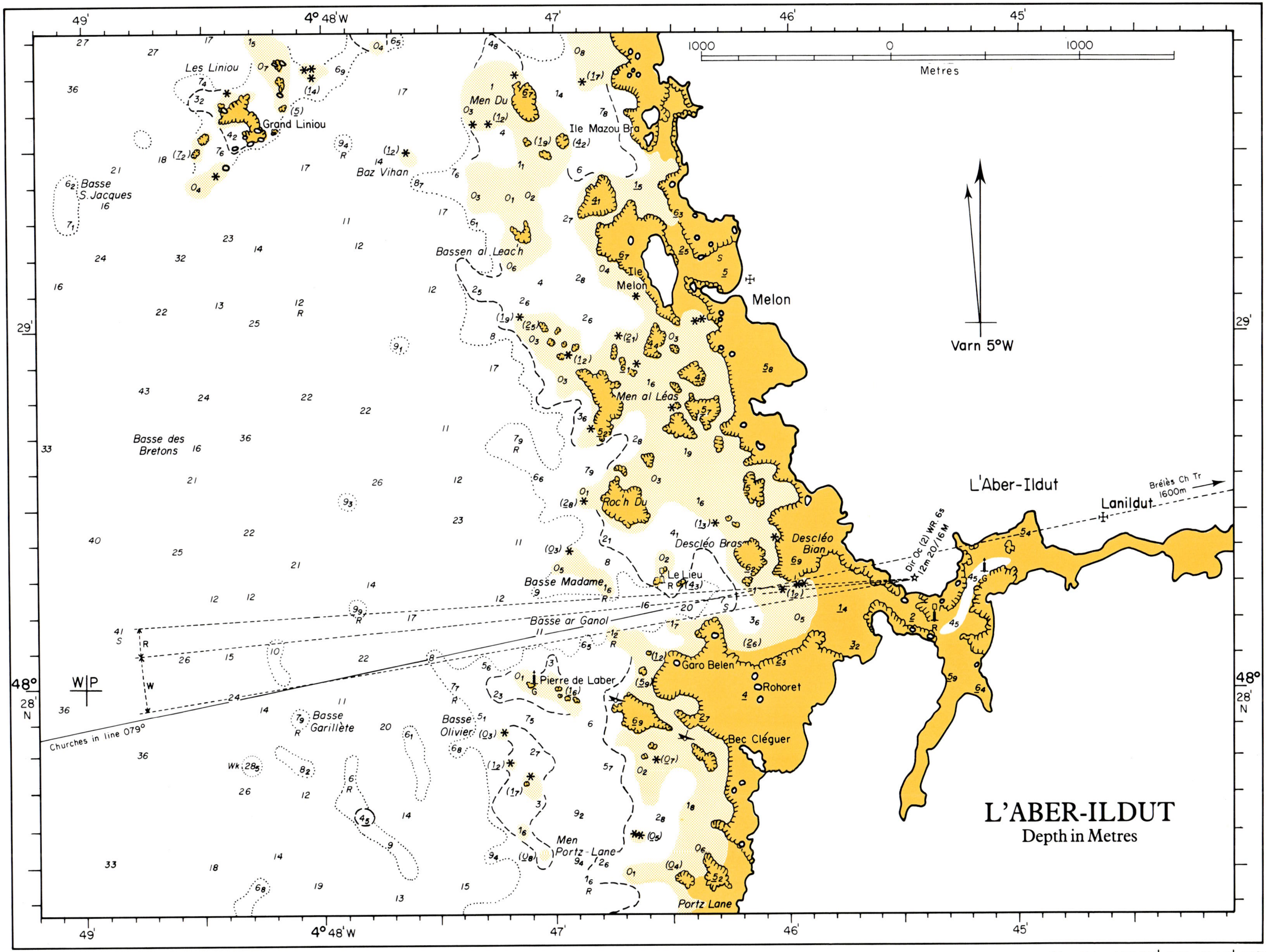

L'ABER-ILDUT

Depth in Metres

Way Pt: 48°28'.0N 4°49'.0W

Le Conquet

See Chart No 354.

TIDE

Time and height differences on Standard Port (for instructions in use, see p. xii)

	Lat	Long	Time Difference HW		LW		Height Difference (metres)			
			0000 and 1200	0600 and 1800	0000 and 1200	0600 and 1800				
BREST See p. 114	48°23′N	4°29′W					7.5	5.9	3.0	1.4
Le Conquet	48°22′N	4°47′W	0000	0000	+0010	0000	−0.3	−0.3	−0.1	0.0

TIDES relative to HW Dover (approx) HW +0520. Stream sets N −0040; S +0525.

GENERAL

Le Conquet provides a deep water (2m) harbour, sheltered from all but W winds and offering some facilities.

APPROACH

The approach is straightforward. Anchor in 2m behind the outer mole, or further up in neaps or with a shallow draft yacht. Do not berth alongside the mole as that is in constant use by the ferry. There are a number of visitors' buoys.

CAUTION

1 The streams run strongly across the entrance.

TELEPHONES

Harbour Master:	(98) 89 00 05
Medical:	(98) 89 01 86 (Doctor)
British Consul:	(99) 46 26 64

VHF

St Mathieu Signal Station:	Ch 16

The chart opposite is based on Chart No 3345 of the Service Hydrographique et Océanographique de la Marine (SHOM), reproduced by permission.

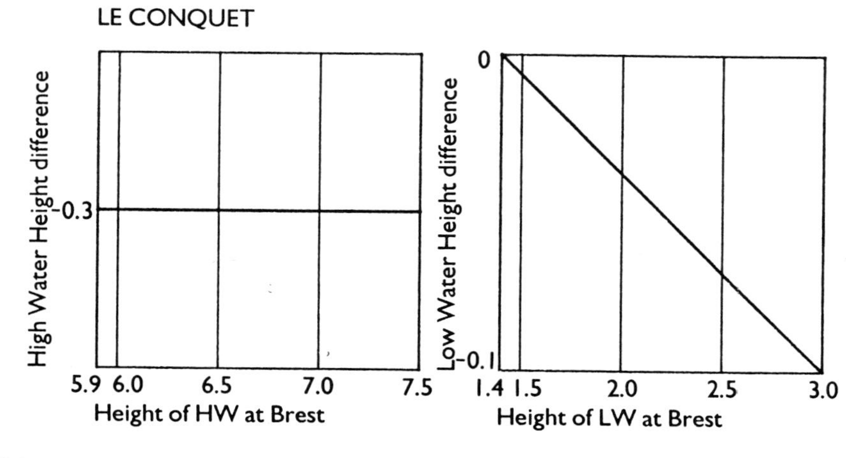

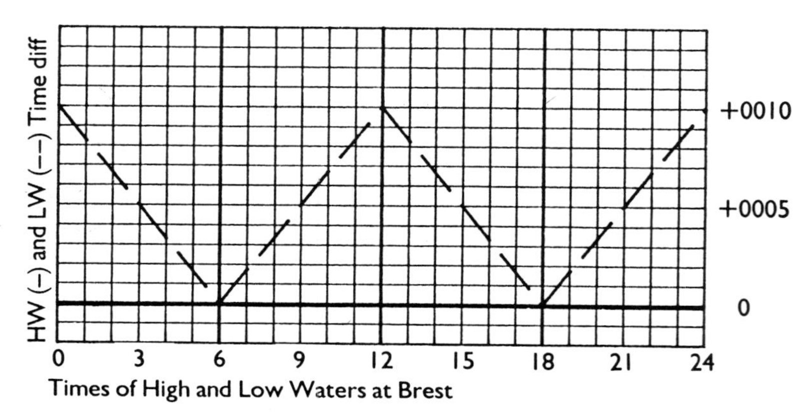

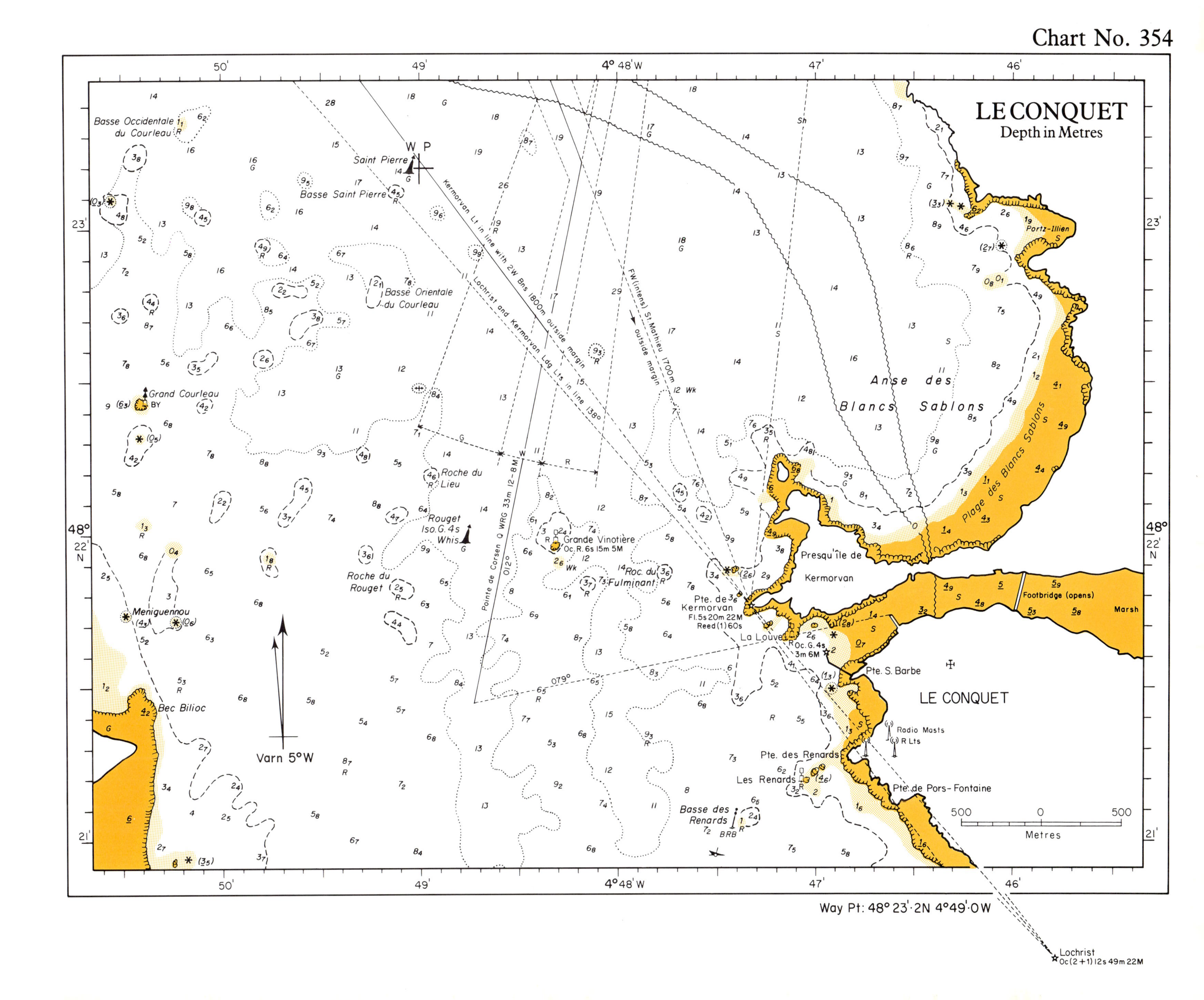
LE CONQUET
Depth in Metres
Portz-Illien
Plage des Blancs Sablons
Anse des Blancs Sablons
Presqu'île de Kermorvan
Marsh
Footbridge (opens)
LE CONQUET
Pte. S. Barbe
Radio Masts
R Lts
Pte. de Pors-Fontaine
Pte. des Renards
Les Renards
Basse des Renards
BRB
Pte. de Kermorvan Fl.5s20m 22M Reed(1)60s
La Louve
Roc. G. 4s 3m 6M
Roc. du Fulminant
Grande Vinotière Oc.R.6s 15m 5M
Wk
FW(intens) St.Mathieu 1700m outside margin
Kermorvan Lt in line with 2W Bns 1800m outside margin
Lochrist and Kermorvan Ldg Lts in line 138°
Pointe de Corsen Q WRG 33m 12-8M
012°
079°
Roche du Lieu
Rouget Iso.G.4s Whis
Roche du Rouget
W P
Saint Pierre
Basse Saint Pierre
Basse Orientale du Courleau
Basse Occidentale du Courleau
Grand Courleau BY
Meriguennou
Bec Bilioc
Varn 5°W
500 0 500
Metres
Way Pt: 48° 23'.2N 4°49'.0W
Lochrist Oc(2+1)12s 49m 22M
50'
49'
4° 48'W
47'
46'
23'
48° 22' N
21'

Île de Molène

See Chart No 355.

TIDE

Time and height differences on Standard Port (for instructions in use, see p. xii)

	Lat	Long	Time Difference HW		LW		Height Difference (metres)			
			0000 and 1200	0600 and 1800	0000 and 1200	0600 and 1800				
BREST See p. 114	48°23′N	4°29′W					7.5	5.9	3.0	1.4
Molène	48°24′N	4°58′W	+0010	+0010	+0015	+0015	0.0	+0.1	−0.1	−0.2

TIDES relative to HW Dover (approx) HW +0515.

Stream sets (NW of Kereon) ENE −0335; WSW +0225.
(between Île de Balance and Île de Bannec) NE −0125; SW +0440.
(in NW channel of Île de Molène – W part) NNE −0120; SSW +0455.
(in NW channel of Île de Molène – E part) ENE −0120; WSW +0455.
(in Chenal des Las) N −0050; S +0255.
(in Passe de la Chimère) N −0120; S +0455.

GENERAL

Île de Molène is suitable only for a visit in good weather. Almost no facilities for yachts.

APPROACH

The harbour is lit by Les Trois Pierres lighthouse. The anchorages, both here and on the NE side of the island, have poor holding. The harbour has many mooring chains lying on the bottom and anchors must be buoyed.

CAUTIONS

1 The island should only be visited for pleasure in settled weather.
2 Wind against tide produces confused and dangerous seas in the passage between Île d'Ouessant and Île de Molène.

The chart opposite is based on Chart No 2694 of the Service Hydrographique et Océanographique de la Marine (SHOM), reproduced by permission.

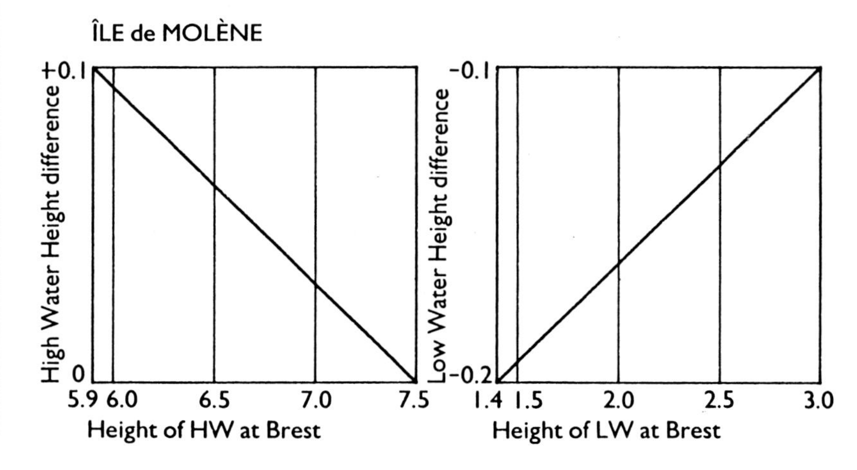

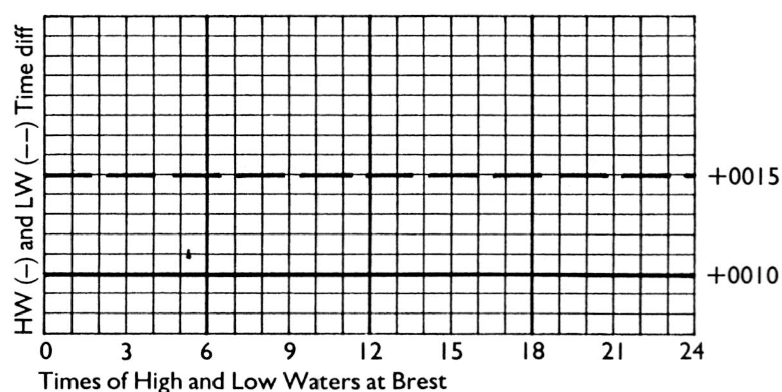

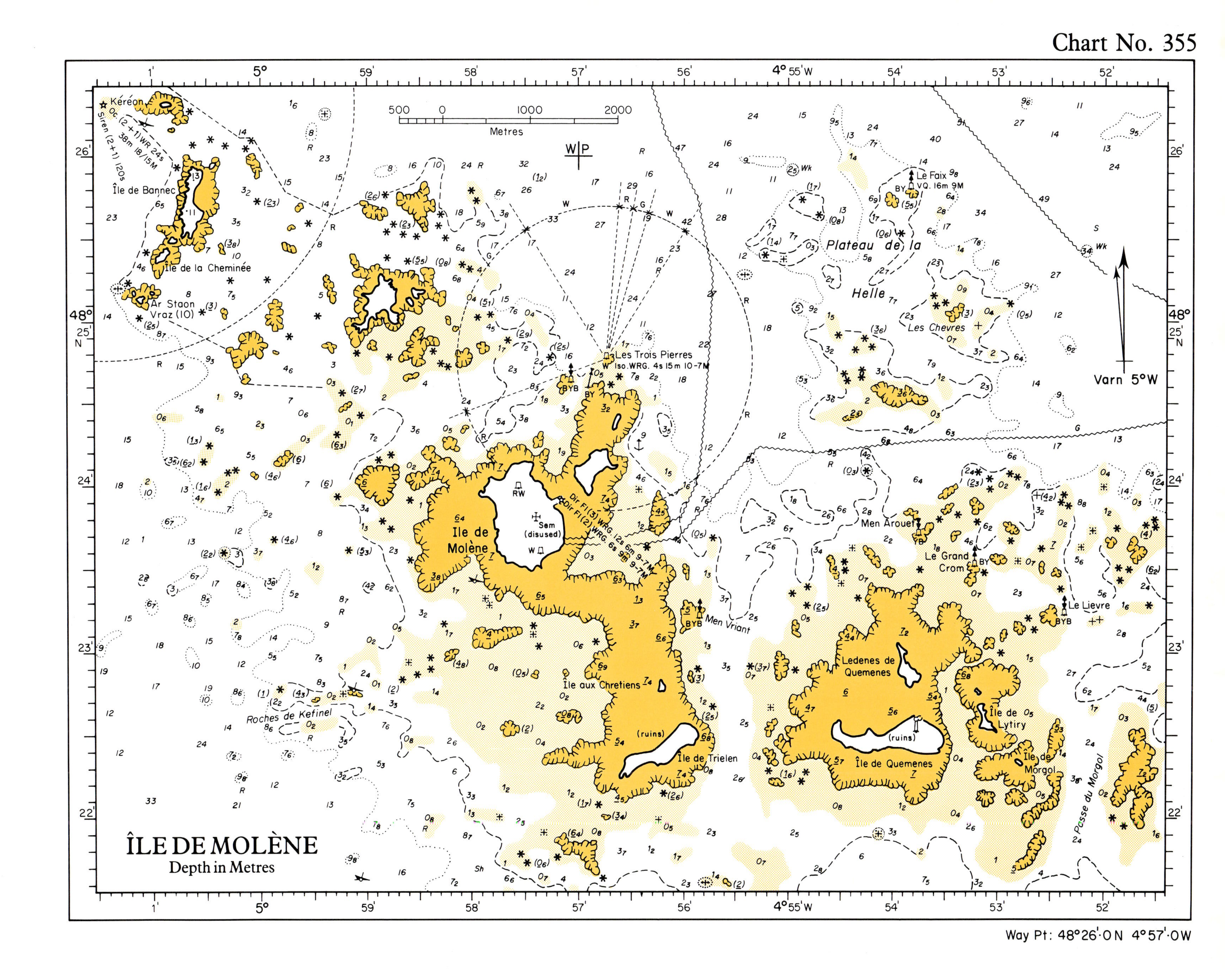

Way Pt: 48°26'·0N 4°57'·0W

Île d'Ouessant (Ushant)

See Chart No 356.

TIDE

Time and height differences on Standard Port (for instructions in use, see p. xii)

	Lat	Long	Time Difference HW		Time Difference LW		Height Difference (metres)			
BREST See p. 114	48°23′N	4°29′W	0000 and 1200	0600 and 1800	0000 and 1200	0600 and 1800	7.5	5.9	3.0	1.4
Baie de Lampaul	48°27′N	5°06′W	0000	+0005	−0005	−0005	0.0	−0.1	0.0	+0.1

TIDES relative to HW Dover (approx) HW +0515.

Stream sets (off Baie du Stiff) NW −0315; SE +0310.
(off NW coast) NE −0040; SW +0225.
(at La Jument) NW −0240; S +0525.
(off S coast) ENE −0335; WSW +0225.

GENERAL

The only harbour on the island is Lampaul which is open to the SW, though sheltered from all other directions. It is used by the Brest and Molène ferries. Space alongside, or a buoy, may be found in the inner harbour, but there is good holding to the S of the beacons marking the harbour entrance. Baie du Stiff should only be used in good weather and Lampaul in offshore winds. Few facilities for yachts.

APPROACH

In reasonable weather and good visibility, neither harbour presents problems, though both have rocks on either side and both have isolated rocks or rocky islands within the entrance. There is a Radio Beacon at Pte de Creac'h (CA 308 kHz 100M).

CAUTIONS

1 The island should only be visited for pleasure in settled weather.
2 Wind against tide produces confused and dangerous seas in the passage between Île d'Ouessant and Île de Molène.

TELEPHONES

Harbour Master:	(98) 89 20 05
Medical:	(98) 89 92 70
British Consul:	(99) 46 16 64

VHF

Ouessant Traffic:	Chs 11; 16; 79

The chart opposite is based on Chart No 2694 of the Service Hydrographique et Océanographique de la Marine (SHOM), reproduced by permission.

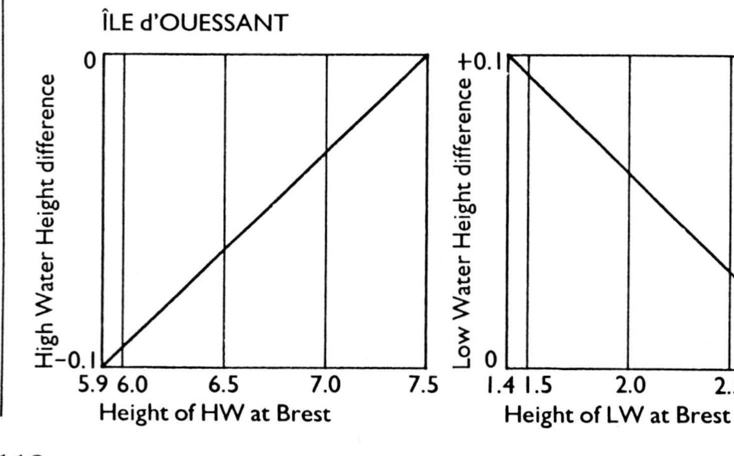

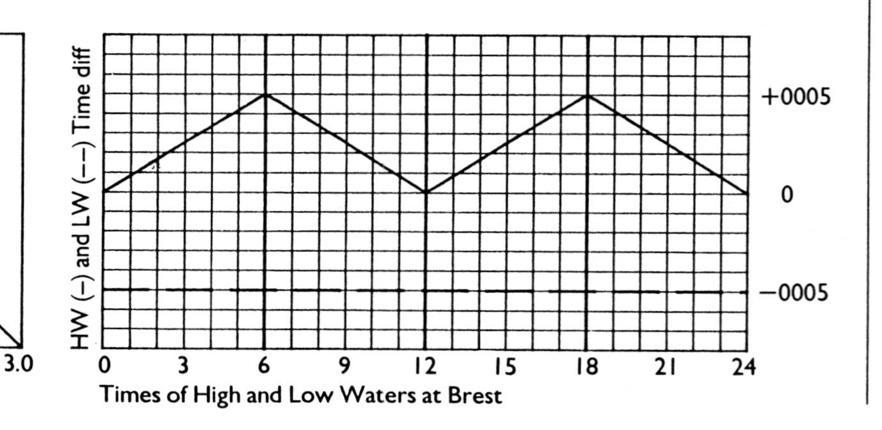

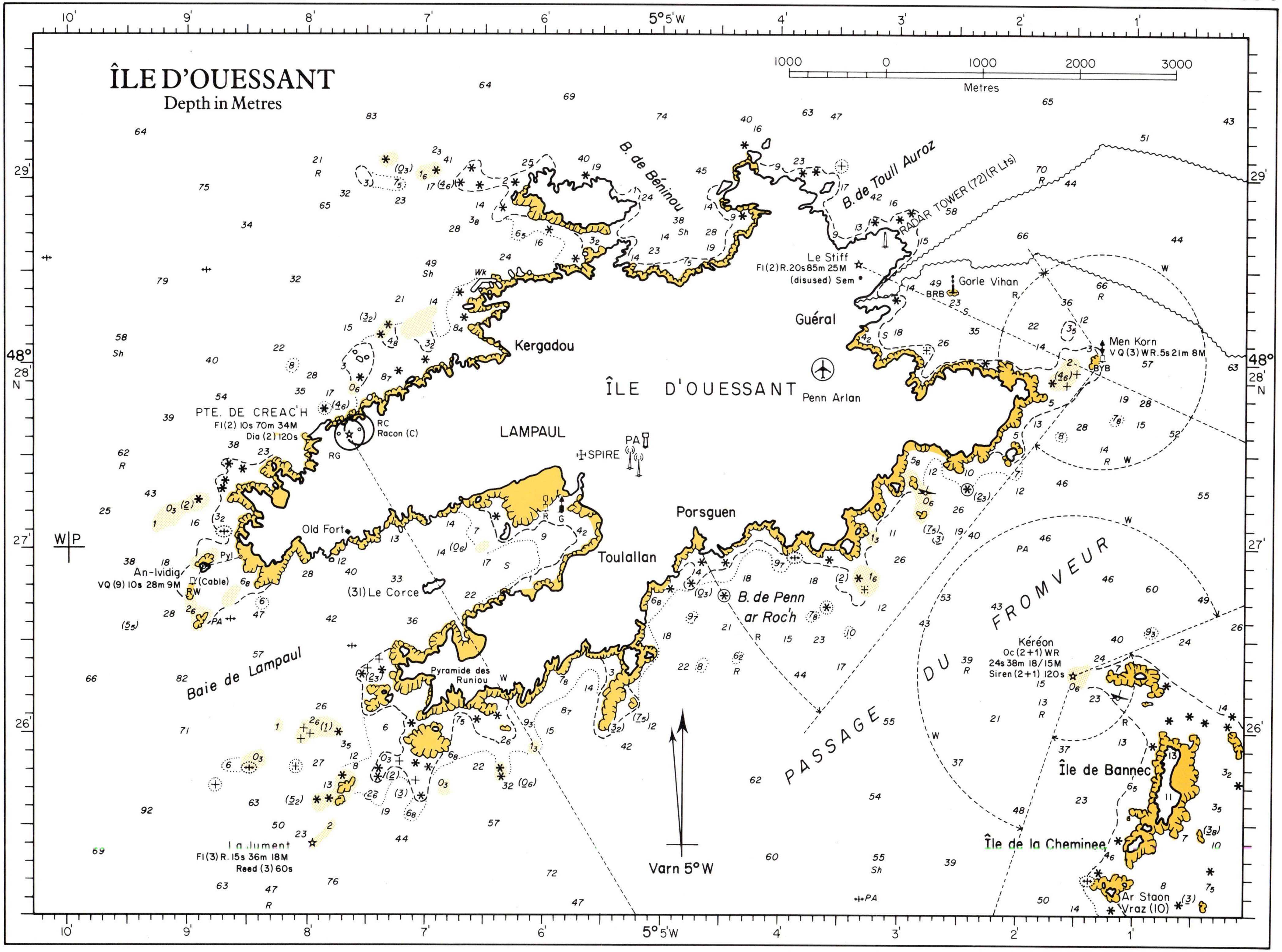
ÎLE D'OUESSANT
Depth in Metres
ÎLE D'OUESSANT
LAMPAUL
Kergadou
Guéral
Penn Arlan
Porsguen
Toulallan
B. de Béninou
B. de Toull Auroz
RADAR TOWER (72)(R Lts)
Le Stiff
Fl(2)R.20s 85m 25M
(disused) Sem
Gorle Vihan
BRB
Men Korn
VQ(3)WR.5s 21m 8M
BYB
PTE. DE CREAC'H
Fl(2) 10s 70m 34M
Dia (2) 120s
RC
Racon (C)
RG
SPIRE
PA
Old Fort
(31) Le Corce
An-Ividig
VQ(9) 10s 28m 9M
(Cable)
RW
Pyl
Baie de Lampaul
Pyramide des Runiou
La Jument
Fl(3)R.15s 36m 18M
Reed (3) 60s
B. de Penn ar Roc'h
PASSAGE DU FROMVEUR
Kéréon
Oc(2+1)WR
24s 38m 18/15M
Siren (2+1) 120s
Île de Bannec
Île de la Cheminee
Ar Staon
Vraz (10)
Varn 5°W
W P
Metres
5°5'W
48° 28' N
Way Pt: 48° 27'·0N 5°10'·0W

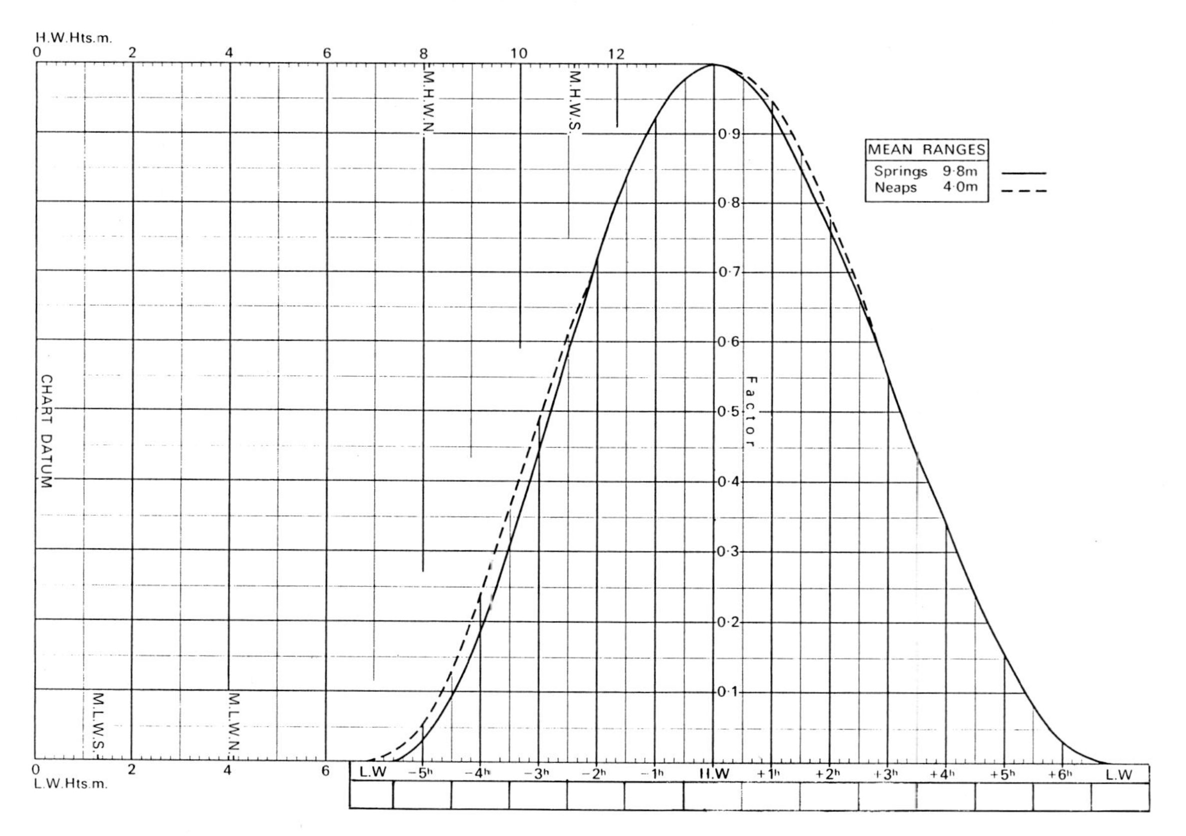
ST HELIER
MEAN SPRING AND NEAP CURVES
Springs occur 2 days after New and Full Moon.
H.W.Hts.m.
M.H.W.N.
M.H.W.S.
MEAN RANGES
Springs 9·8m
Neaps 4·0m
CHART DATUM
Factor
M.L.W.S.
M.L.W.N.
L.W.Hts.m.
L.W -5h -4h -3h -2h -1h H.W +1h +2h +3h +4h +5h +6h L.W

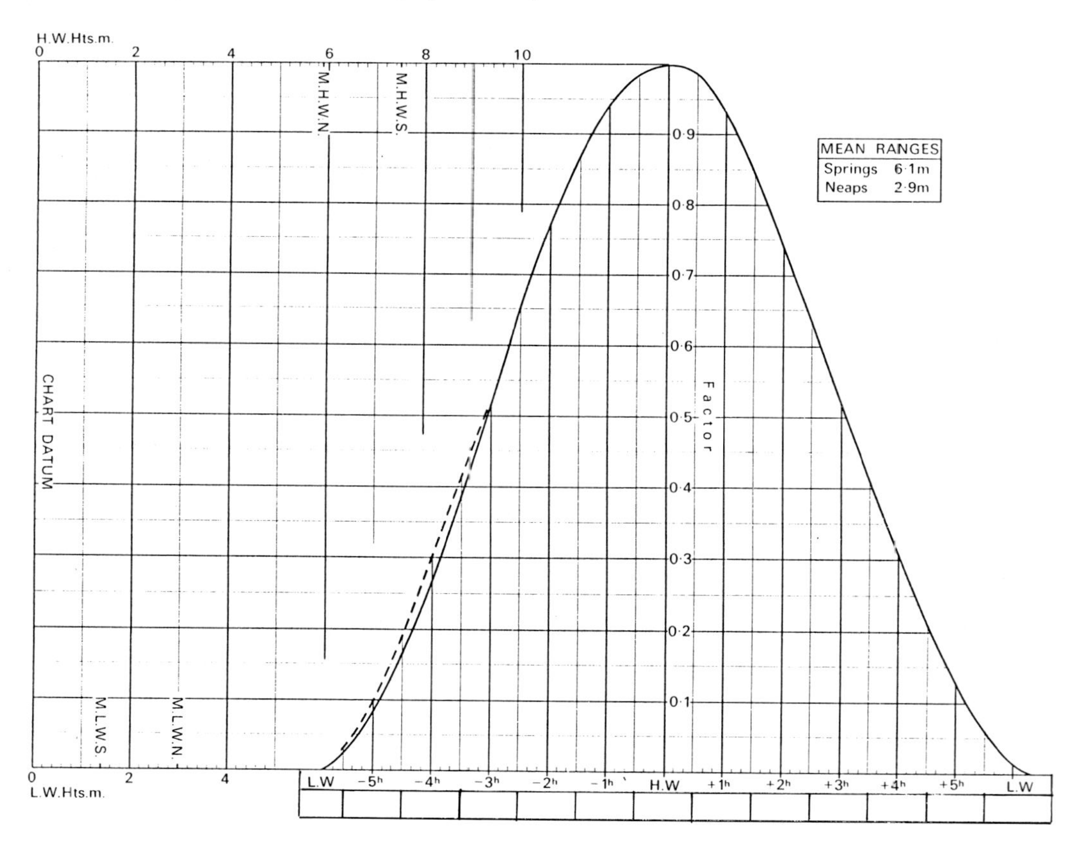
BREST
MEAN SPRING AND NEAP CURVES
Springs occur 2 days after New and Full Moon.
H.W.Hts.m.
M.H.W.N.
M.H.W.S.
MEAN RANGES
Springs 6·1m
Neaps 2·9m
CHART DATUM
Factor
M.L.W.S.
M.L.W.N.
L.W.Hts.m.
L.W -5h -4h -3h -2h -1h H.W +1h +2h +3h +4h +5h L.W

DOVER

51°07′N 1°19′E

Times and heights (**in metres**) of high and low water 1989
Time: GMT. For BST, ADD ONE HOUR in the shaded area
High Water, full and change, 1110

JANUARY

	Time	M		Time	M
1 SU	0452 1156 1727	5.4 2.0 5.1	16 M	0519 1243 1804	5.9 1.7 5.5
2 M	0015 0554 1257 1834	2.3 5.3 2.1 5.1	17 TU	0109 0631 1359 1923	2.0 5.7 1.8 5.4
3 TU	0124 0659 1406 1935	2.3 5.3 2.1 5.3	18 W	0230 0748 1519 2039	2.0 5.7 1.7 5.6
4 W	0242 0758 1518 2030	2.2 5.5 1.9 5.5	19 TH	0348 0900 1630 2139	1.8 5.8 1.6 5.8
5 TH	0353 0850 1621 2119	1.9 5.7 1.6 5.8	20 F	0451 0957 1727 2226	1.5 6.0 1.4 6.1
6 F	0451 0938 1715 2206	1.6 5.9 1.4 6.1	21 SA	0544 1045 1817 ○ 2308	1.3 6.1 1.3 6.3
7 SA	0542 1023 1804 ● 2251	1.3 6.2 1.2 6.3	22 SU	0629 1127 1857 2346	1.1 6.2 1.2 6.5
8 SU	0628 1108 1848 2334	1.1 6.4 1.1 6.5	23 M	0709 1205 1930	1.0 6.3 1.1
9 M	0710 1153 1928	0.9 6.5 1.0	24 TU	0022 0744 1239 1958	6.5 0.9 6.3 1.1
10 TU	0019 0751 1241 2008	6.6 0.8 6.5 0.9	25 W	0056 0815 1312 2025	6.5 1.0 6.2 1.2
11 W	0104 0832 1328 2049	6.7 0.8 6.4 0.9	26 TH	0128 0846 1341 2051	6.4 1.0 6.1 1.3
12 TH	0151 0914 1416 2129	6.6 0.8 6.3 1.0	27 F	0158 0915 1411 2119	6.3 1.2 6.0 1.4
13 F	0236 0957 1505 2213	6.5 0.9 6.1 1.2	28 SA	0229 0946 1443 2150	6.1 1.4 5.8 1.6
14 SA	0324 1044 1556 2259	6.4 1.1 5.9 1.5	29 SU	0301 1017 1519 2224	5.9 1.6 5.6 1.9
15 SU	0417 1137 1655 2357	6.2 1.4 5.7 1.8	30 M	0341 1057 1607 2311	5.7 1.9 5.3 2.2
			31 TU	0434 1151 1713	5.4 2.2 5.1

FEBRUARY

	Time	M		Time	M
1 W	0018 0549 1307 1836	2.4 5.1 2.3 5.0	16 TH	0215 0742 1514 2033	2.1 5.3 2.0 5.3
2 TH	0148 0713 1436 1958	2.4 5.1 2.1 5.2	17 F	0345 0901 1627 2132	1.8 5.5 1.7 5.7
3 F	0321 0825 1555 2101	2.1 5.4 1.8 5.6	18 SA	0449 0956 1725 2216	1.4 5.8 1.4 6.0
4 SA	0430 0922 1657 2152	1.6 5.8 1.4 6.0	19 SU	0540 1038 1810 2254	1.2 6.0 1.2 6.3
5 SU	0526 1012 1750 2238	1.2 6.1 1.1 6.3	20 M	0621 1115 1845 ○ 2327	1.0 6.2 1.1 6.4
6 M	0615 1057 1839 ● 2322	0.9 6.4 0.9 6.6	21 TU	0655 1147 1912	0.9 6.3 1.0
7 TU	0702 1142 1923	0.6 6.6 0.7	22 W	0000 0723 1217 1934	6.5 0.8 6.3 1.0
8 W	0004 0744 1227 2001	6.8 0.5 6.7 0.6	23 TH	0031 0749 1242 1957	6.6 0.8 6.3 1.0
9 TH	0046 0823 1310 2034	6.9 0.4 6.7 0.6	24 F	0057 0816 1306 2022	6.5 0.9 6.3 1.0
10 F	0128 0901 1354 2111	6.9 0.5 6.5 0.7	25 SA	0121 0844 1330 2049	6.4 1.0 6.2 1.2
11 SA	0211 0938 1437 2148	6.8 0.7 6.3 1.0	26 SU	0144 0911 1357 2115	6.3 1.2 6.0 1.4
12 SU	0256 1017 1524 2228	6.6 1.0 6.0 1.4	27 M	0211 0939 1429 2145	6.1 1.5 5.8 1.7
13 M	0345 1104 1619 2320	6.2 1.4 5.7 1.8	28 TU	0244 1010 1510 2223	5.8 1.8 5.5 2.1
14 TU	0445 1207 1730	5.8 1.9 5.3			
15 W	0035 0605 1335 1904	2.1 5.4 2.1 5.1			

MARCH

	Time	M		Time	M
1 W	0332 1101 1610 2327	5.5 2.1 5.1 2.4	16 TH	0014 0551 1317 1849	2.2 5.1 2.3 5.0
2 TH	0449 1219 1746	5.1 2.4 4.9	17 F	0201 0740 1500 2016	2.2 5.1 2.1 5.3
3 F	0103 0639 1359 1931	2.4 4.9 2.3 5.0	18 SA	0331 0853 1609 2112	1.7 5.4 1.7 5.7
4 SA	0250 0808 1529 2043	2.1 5.3 1.8 5.5	19 SU	0430 0942 1701 2153	1.3 5.7 1.4 6.0
5 SU	0406 0908 1634 2134	1.5 5.7 1.4 6.0	20 M	0518 1020 1743 2230	1.1 6.0 1.2 6.2
6 M	0504 0955 1729 2219	1.0 6.1 1.0 6.4	21 TU	0556 1051 1817 2302	0.9 6.1 1.1 6.4
7 TU	0556 1038 1819 ● 2259	0.7 6.5 0.7 6.7	22 W	0627 1119 1842 ○ 2333	0.9 6.3 1.0 6.5
8 W	0643 1120 1903 2342	0.4 6.7 0.5 7.0	23 TH	0655 1146 1904	0.8 6.3 0.9
9 TH	0727 1203 1940	0.3 6.8 0.4	24 F	0000 0720 1208 1928	6.5 0.8 6.4 0.9
10 F	0021 0805 1243 2013	7.1 0.2 6.8 0.5	25 SA	0022 0747 1231 1955	6.5 0.8 6.3 1.0
11 SA	0102 0840 1326 2049	7.0 0.4 6.6 0.6	26 SU	0043 0815 1256 2023	6.4 1.0 6.3 1.1
12 SU	0144 0917 1409 2125	6.8 0.6 6.4 0.9	27 M	0107 0843 1323 2053	6.3 1.2 6.1 1.3
13 M	0227 0955 1456 2206	6.5 1.1 6.0 1.3	28 TU	0134 0911 1355 2122	6.1 1.4 6.0 1.6
14 TU	0318 1038 1552 2257	6.1 1.6 5.6 1.8	29 W	0209 0945 1436 2202	5.9 1.7 5.7 1.9
15 W	0421 1142 1705	5.5 2.1 5.2	30 TH	0257 1034 1536 2305	5.5 2.1 5.3 2.2
			31 F	0417 1153 1716	5.1 2.3 5.0

APRIL

	Time	M		Time	M
1 SA	0036 0621 1331 1907	2.2 5.0 2.2 5.1	16 SU	0254 0822 1529 2037	1.7 5.4 1.8 5.7
2 SU	0219 0749 1458 2018	1.9 5.4 1.8 5.6	17 M	0352 0908 1620 2119	1.4 5.7 1.5 5.9
3 M	0334 0846 1603 2108	1.4 5.8 1.3 6.1	18 TU	0438 0946 1701 2156	1.2 5.9 1.3 6.1
4 TU	0433 0932 1658 2150	0.9 6.2 0.9 6.5	19 W	0516 1017 1736 2230	1.1 6.0 1.2 6.3
5 W	0526 1013 1749 2233	0.6 6.5 0.7 6.8	20 TH	0550 1045 1804 2259	1.0 6.2 1.1 6.4
6 TH	0615 1055 1834 ● 2313	0.4 6.7 0.5 7.0	21 F	0621 1111 1832 ○ 2326	0.9 6.3 1.0 6.4
7 F	0700 1137 1912 2354	0.3 6.8 0.5 7.1	22 SA	0650 1137 1902 2350	0.9 6.3 1.0 6.4
8 SA	0740 1219 1949	0.3 6.8 0.5	23 SU	0720 1203 1933	0.9 6.3 1.0
9 SU	0036 0819 1302 2029	7.0 0.5 6.6 0.7	24 M	0015 0751 1231 2005	6.3 1.0 6.3 1.1
10 M	0120 0857 1348 2108	6.7 0.8 6.3 1.0	25 TU	0042 0822 1302 2037	6.2 1.2 6.1 1.3
11 TU	0206 0936 1436 2152	6.4 1.2 6.0 1.3	26 W	0114 0856 1338 2114	6.0 1.4 6.0 1.5
12 W	0300 1023 1532 2244	5.9 1.7 5.6 1.8	27 TH	0154 0935 1425 2159	5.8 1.7 5.7 1.7
13 TH	0406 1123 1642 2356	5.4 2.1 5.2 2.1	28 F	0249 1026 1531 2258	5.5 1.9 5.4 1.9
14 F	0537 1250 1819	5.0 2.3 5.1	29 SA	0416 1136 1702	5.2 2.1 5.2
15 SA	0131 0716 1422 1941	2.1 5.1 2.1 5.3	30 SU	0018 0603 1303 1835	1.9 5.2 2.0 5.4

HEIGHT IN METRES

DOVER

51°07′N 1°19′E

Times and heights (**in metres**) of high and low water 1989
Time: GMT. For BST, ADD ONE HOUR in the shaded area
High Water, full and change, 1110

MAY

Day	Time	M	Day	Time	M
1 M	0145	1.6	16 TU	0257	1.7
	0720	5.5		0823	5.5
	1422	1.6		1524	1.8
	1942	5.8		2036	5.8
2 TU	0257	1.3	17 W	0348	1.5
	0815	5.9		0904	5.7
	1525	1.3		1610	1.6
	2034	6.2		2117	5.9
3 W	0357	0.9	18 TH	0430	1.4
	0901	6.2		0938	5.9
	1620	1.0		1651	1.4
	2119	6.5		2152	6.1
4 TH	0451	0.7	19 F	0509	1.2
	0946	6.4		1009	6.0
	1712	0.9		1729	1.3
	2204	6.7		2224	6.2
5 F	0543	0.6	20 SA	0547	1.1
	1030	6.6		1040	6.2
	1800	0.7		1805	1.2
	● 2248	6.9		O 2254	6.2
6 SA	0631	0.5	21 SU	0624	1.1
	1115	6.7		1111	6.2
	1845	0.6		1842	1.1
	2333	6.9		2325	6.2
7 SU	0716	0.6	22 M	0659	1.1
	1201	6.6		1144	6.3
	1930	0.7		1917	1.1
				2357	6.2
8 M	0018	6.7	23 TU	0734	1.1
	0759	0.7		1219	6.2
	1248	6.5		1952	1.1
	2013	0.8			
9 TU	0106	6.5	24 W	0032	6.1
	0842	1.0		0809	1.2
	1334	6.3		1259	6.2
	2057	1.0		2030	1.2
10 W	0155	6.2	25 TH	0112	6.0
	0924	1.3		0847	1.3
	1422	6.0		1341	6.0
	2141	1.3		2111	1.3
11 TH	0247	5.8	26 F	0159	5.8
	1009	1.7		0931	1.4
	1514	5.8		1433	5.9
	2231	1.6		2156	1.4
12 F	0349	5.4	27 SA	0301	5.6
	1101	2.0		1020	1.6
	1616	5.5		1534	5.7
	2330	1.9		2251	1.5
13 SA	0505	5.2	28 SU	0416	5.5
	1207	2.2		1120	1.7
	1733	5.3		1644	5.7
				2357	1.5
14 SU	0042	1.9	29 M	0532	5.5
	0628	5.2		1231	1.7
	1321	2.2		1756	5.7
	1849	5.4			
15 M	0155	1.8	30 TU	0110	1.4
	0733	5.3		0641	5.7
	1429	2.0		1342	1.6
	1948	5.6		1902	5.9
			31 W	0219	1.3
				0740	5.9
				1446	1.4
				1958	6.1

JUNE

Day	Time	M	Day	Time	M
1 TH	0321	1.1	16 F	0341	1.7
	0832	6.1		0857	5.6
	1545	1.3		1609	1.7
	2050	6.4		2112	5.8
2 F	0420	1.0	17 SA	0433	1.5
	0922	6.2		0936	5.8
	1641	1.1		1658	1.5
	2141	6.5		2152	5.9
3 SA	0516	0.9	18 SU	0519	1.3
	1013	6.4		1014	6.0
	1736	1.0		1744	1.3
	● 2231	6.6		2230	6.0
4 SU	0610	0.8	19 M	0604	1.2
	1104	6.5		1054	6.2
	1827	0.9		1827	1.2
	2322	6.6		O 2309	6.1
5 M	0700	0.8	20 TU	0645	1.1
	1151	6.5		1133	6.3
	1916	0.8		1906	1.1
				2350	6.2
6 TU	0011	6.5	21 W	0723	1.1
	0745	0.9		1215	6.3
	1238	6.4		1945	1.0
	2002	0.9			
7 W	0057	6.3	22 TH	0032	6.2
	0829	1.1		0801	1.1
	1321	6.3		1259	6.3
	2046	1.0		2025	1.0
8 TH	0144	6.1	23 F	0119	6.1
	0910	1.3		0840	1.1
	1406	6.2		1345	6.3
	2128	1.2		2105	1.0
9 F	0230	5.9	24 SA	0209	6.0
	0949	1.5		0922	1.2
	1451	6.0		1433	6.2
	2210	1.4		2149	1.0
10 SA	0321	5.6	25 SU	0301	5.9
	1030	1.8		1007	1.3
	1542	5.8		1522	6.1
	2255	1.6		2237	1.2
11 SU	0419	5.4	26 M	0356	5.8
	1113	2.0		1058	1.4
	1638	5.6		1616	6.1
	2344	1.8		2330	1.3
12 M	0522	5.2	27 TU	0454	5.7
	1205	2.1		1156	1.6
	1743	5.4		1716	6.0
13 TU	0041	1.9	28 W	0034	1.4
	0627	5.2		0558	5.7
	1304	2.2		1302	1.7
	1846	5.4		1821	5.9
14 W	0141	1.9	29 TH	0142	1.4
	0724	5.3		0704	5.7
	1408	2.1		1411	1.6
	1942	5.5		1928	6.0
15 TH	0243	1.8	30 F	0253	1.4
	0813	5.4		0811	5.8
	1511	1.9		1519	1.5
	2030	5.6		2032	6.0

JULY

Day	Time	M	Day	Time	M
1 SA	0400	1.3	16 SU	0400	1.8
	0912	5.9		0911	5.6
	1626	1.4		1633	1.7
	2131	6.2		2129	5.7
2 SU	0504	1.2	17 M	0457	1.5
	1009	6.1		0956	5.9
	1725	1.2		1725	1.4
	2227	6.3		2213	6.0
3 M	0601	1.1	18 TU	0547	1.3
	1058	6.3		1038	6.1
	1819	1.0		1812	1.1
	● 2318	6.4		O 2257	6.2
4 TU	0652	1.0	19 W	0634	1.1
	1144	6.4		1120	6.4
	1907	0.9		1856	0.9
				2340	6.3
5 W	0004	6.3	20 TH	0714	1.0
	0737	1.0		1204	6.5
	1225	6.5		1937	0.8
	1951	0.9			
6 TH	0048	6.3	21 F	0024	6.4
	0816	1.1		0754	0.9
	1304	6.4		1246	6.6
	2030	0.9		2016	0.7
7 F	0126	6.2	22 SA	0109	6.4
	0850	1.2		0830	0.8
	1344	6.4		1330	6.6
	2107	1.0		2056	0.7
8 SA	0205	6.0	23 SU	0155	6.4
	0921	1.3		0908	0.9
	1422	6.2		1412	6.6
	2142	1.2		2135	0.8
9 SU	0244	5.8	24 M	0239	6.2
	0952	1.5		0949	1.0
	1501	6.0		1457	6.5
	2216	1.4		2217	0.9
10 M	0325	5.6	25 TU	0325	6.1
	1024	1.7		1031	1.3
	1543	5.8		1545	6.3
	2252	1.6		2304	1.2
11 TU	0410	5.4	26 W	0419	5.8
	1102	1.9		1120	1.6
	1631	5.6		1641	6.0
	2334	1.9			
12 W	0505	5.2	27 TH	0000	1.5
	1150	2.2		0522	5.6
	1730	5.3		1225	1.8
				1750	5.8
13 TH	0028	2.0	28 F	0114	1.7
	0608	5.1		0639	5.4
	1253	2.3		1347	1.9
	1836	5.2		1910	5.6
14 F	0135	2.1	29 SA	0237	1.8
	0717	5.1		0805	5.5
	1411	2.3		1511	1.8
	1942	5.3		2030	5.7
15 SA	0251	2.0	30 SU	0357	1.6
	0819	5.3		0914	5.8
	1531	2.0		1624	1.5
	2040	5.5		2135	5.9
			31 M	0505	1.4
				1006	6.0
				1725	1.2
				2227	6.1

AUGUST

Day	Time	M	Day	Time	M
1 TU	0601	1.2	16 W	0529	1.2
	1051	6.3		1021	6.2
	1817	1.0		1754	1.0
	● 2312	6.2		2240	6.3
2 W	0649	1.1	17 TH	0617	1.0
	1130	6.5		1102	6.6
	1900	0.9		1842	0.7
	2351	6.3		O 2320	6.5
3 TH	0726	1.0	18 F	0702	0.8
	1207	6.6		1142	6.8
	1937	0.8		1924	0.5
4 F	0028	6.3	19 SA	0003	6.7
	0755	1.0		0740	0.7
	1242	6.6		1222	6.9
	2009	0.9		2002	0.5
5 SA	0100	6.3	20 SU	0043	6.7
	0822	1.1		0813	0.7
	1314	6.5		1303	6.9
	2039	0.9		2039	0.5
6 SU	0130	6.2	21 M	0126	6.6
	0847	1.2		0849	0.8
	1347	6.4		1344	6.8
	2108	1.1		2115	0.7
7 M	0159	6.1	22 TU	0208	6.4
	0912	1.3		0925	1.0
	1416	6.2		1426	6.6
	2136	1.3		2155	0.9
8 TU	0230	5.9	23 W	0254	6.2
	0941	1.6		1006	1.3
	1447	6.0		1515	6.3
	2207	1.6		2237	1.3
9 W	0305	5.6	24 TH	0348	5.8
	1013	1.8		1054	1.7
	1524	5.7		1613	5.9
	2242	1.9		2334	1.8
10 TH	0349	5.4	25 F	0454	5.4
	1055	2.2		1201	2.0
	1614	5.4		1729	5.5
	2332	2.2			
11 F	0451	5.1	26 SA	0057	2.1
	1156	2.4		0627	5.2
	1727	5.0		1335	2.1
				1912	5.3
12 SA	0041	2.3	27 SU	0236	2.0
	0617	4.9		0808	5.4
	1320	2.5		1512	1.8
	1859	5.0		2040	5.6
13 SU	0208	2.3	28 M	0357	1.7
	0747	5.1		0911	5.8
	1456	2.2		1623	1.5
	2016	5.3		2139	5.8
14 M	0332	1.9	29 TU	0501	1.4
	0850	5.5		0956	6.1
	1609	1.7		1719	1.1
	2112	5.6		2221	6.1
15 TU	0435	1.6	30 W	0551	1.2
	0939	5.9		1034	6.3
	1705	1.3		1805	1.0
	2157	6.0		2258	6.2
			31 TH	0632	1.1
				1109	6.5
				1842	0.9
				● 2330	6.3

HEIGHT IN METRES

DOVER

51°07′N 1°19′E

Times and heights (**in metres**) of high and low water 1989
Time: GMT. For BST, ADD ONE HOUR in the shaded area
High Water, full and change, 1110

SEPTEMBER

	Time	M		Time	M
1 F	0702	1.1	16 SA	0636	0.7
	1143	6.6		1115	7.0
	1912	0.8		1900	0.5
				2334	6.8
2 SA	0000	6.4	17 SU	0714	0.6
	0726	1.1		1153	7.1
	1214	6.7		1940	0.4
	1938	0.9			
3 SU	0028	6.4	18 M	0014	6.8
	0747	1.1		0749	0.7
	1242	6.6		1232	7.1
	2005	0.9		2016	0.5
4 M	0052	6.3	19 TU	0056	6.7
	0812	1.2		0825	0.8
	1306	6.5		1314	6.9
	2033	1.1		2054	0.7
5 TU	0116	6.2	20 W	0140	6.5
	0837	1.3		0904	1.0
	1330	6.3		1359	6.6
	2100	1.3		2134	1.1
6 W	0142	6.1	21 TH	0229	6.2
	0905	1.5		0946	1.4
	1355	6.1		1450	6.2
	2128	1.6		2219	1.6
7 TH	0213	5.9	22 F	0324	5.8
	0935	1.8		1037	1.8
	1427	5.8		1552	5.7
	2200	1.9		2319	2.0
8 F	0251	5.6	23 SA	0434	5.4
	1014	2.2		1150	2.1
	1510	5.4		1718	5.2
	2247	2.2			
9 SA	0348	5.2	24 SU	0048	2.3
	1115	2.5		0615	5.1
	1624	5.0		1330	2.2
				1917	5.2
10 SU	0000	2.5	25 M	0227	2.1
	0525	4.9		0754	5.4
	1241	2.6		1501	1.8
	1825	4.9		2034	5.5
11 M	0133	2.4	26 TU	0341	1.7
	0716	5.0		0853	5.8
	1423	2.3		1606	1.4
	1957	5.2		2125	5.9
12 TU	0301	2.0	27 W	0437	1.4
	0827	5.5		0934	6.1
	1541	1.7		1655	1.1
	2053	5.7		2202	6.1
13 W	0407	1.6	28 TH	0522	1.3
	0915	6.0		1010	6.4
	1638	1.2		1737	1.0
	2136	6.1		2233	6.2
14 TH	0502	1.2	29 F	0558	1.2
	0956	6.4		1042	6.5
	1729	0.8		1810	1.0
	2216	6.5		● 2301	6.4
15 F	0551	0.9	30 SA	0625	1.1
	1035	6.7		1113	6.6
	1817	0.6		1838	1.0
	O 2255	6.7		2329	6.4

OCTOBER

	Time	M		Time	M
1 SU	0649	1.1	16 M	0643	0.7
	1143	6.6		1125	7.2
	1904	1.0		1912	0.5
	2354	6.5		2349	6.9
2 M	0713	1.1	17 TU	0723	0.7
	1207	6.6		1207	7.1
	1931	1.0		1952	0.7
3 TU	0017	6.4	18 W	0034	6.7
	0740	1.2		0805	0.9
	1229	6.5		1252	6.9
	1959	1.1		2034	0.9
4 W	0041	6.3	19 TH	0120	6.5
	0809	1.3		0847	1.1
	1252	6.3		1340	6.5
	2029	1.3		2118	1.3
5 TH	0107	6.2	20 F	0211	6.2
	0840	1.6		0934	1.4
	1319	6.1		1433	6.1
	2058	1.6		2206	1.7
6 F	0138	6.0	21 SA	0307	5.8
	0911	1.8		1027	1.8
	1351	5.9		1538	5.6
	2132	1.9		2306	2.1
7 SA	0216	5.7	22 SU	0414	5.5
	0952	2.1		1137	2.1
	1433	5.5		1705	5.2
	2219	2.2			
8 SU	0310	5.3	23 M	0028	2.3
	1049	2.4		0546	5.3
	1545	5.1		1306	2.1
	2329	2.5		1852	5.2
9 M	0445	5.0	24 TU	0154	2.2
	1212	2.4		0714	5.4
	1757	4.9		1427	1.8
				2001	5.5
10 TU	0100	2.4	25 W	0303	1.9
	0641	5.1		0815	5.8
	1348	2.1		1528	1.5
	1928	5.3		2050	5.8
11 W	0226	2.0	26 TH	0356	1.6
	0754	5.6		0900	6.0
	1504	1.6		1617	1.3
	2025	5.8		2128	6.0
12 TH	0332	1.6	27 F	0440	1.5
	0843	6.1		0938	6.2
	1603	1.2		1658	1.2
	2107	6.2		2200	6.1
13 F	0427	1.2	28 SA	0516	1.4
	0925	6.5		1012	6.4
	1657	0.8		1732	1.2
	2146	6.5		2230	6.3
14 SA	0516	0.9	29 SU	0547	1.3
	1004	6.8		1042	6.5
	1746	0.6		1803	1.1
	O 2226	6.8		● 2258	6.4
15 SU	0601	0.8	30 M	0615	1.2
	1045	7.1		1111	6.5
	1831	0.5		1832	1.1
	2306	6.9		2323	6.4
			31 TU	0645	1.2
				1136	6.5
				1903	1.1
				2350	6.4

NOVEMBER

	Time	M		Time	M
1 W	0716	1.3	16 TH	0019	6.7
	1201	6.4		0751	0.9
	1934	1.2		1238	6.7
				2020	1.0
2 TH	0017	6.3	17 F	0109	6.5
	0749	1.4		0837	1.1
	1228	6.3		1328	6.4
	2005	1.4		2105	1.3
3 F	0048	6.2	18 SA	0158	6.3
	0823	1.5		0924	1.3
	1257	6.1		1422	6.0
	2039	1.6		2153	1.7
4 SA	0121	6.1	19 SU	0250	6.0
	0858	1.7		1014	1.6
	1334	5.9		1521	5.7
	2117	1.8		2245	2.0
5 SU	0202	5.8	20 M	0348	5.7
	0941	1.9		1113	1.9
	1422	5.6		1630	5.4
	2203	2.1		2349	2.2
6 M	0258	5.5	21 TU	0457	5.5
	1035	2.1		1222	2.0
	1534	5.2		1754	5.2
	2306	2.3			
7 TU	0421	5.3	22 W	0057	2.3
	1146	2.1		0615	5.5
	1720	5.2		1333	1.9
				1906	5.3
8 W	0025	2.2	23 TH	0205	2.2
	0556	5.4		0723	5.6
	1309	1.9		1434	1.8
	1845	5.4		2001	5.5
9 TH	0145	2.0	24 F	0303	2.0
	0707	5.7		0815	5.8
	1423	1.6		1528	1.7
	1944	5.8		2047	5.7
10 F	0253	1.6	25 SA	0352	1.8
	0804	6.1		0900	6.0
	1525	1.2		1614	1.5
	2032	6.1		2125	5.9
11 SA	0349	1.3	26 SU	0434	1.7
	0850	6.5		0938	6.1
	1620	1.0		1654	1.4
	2117	6.4		2159	6.1
12 SU	0440	1.1	27 M	0513	1.5
	0934	6.8		1013	6.2
	1712	0.8		1730	1.3
	2200	6.7		2230	6.2
13 M	0529	0.9	28 TU	0550	1.4
	1019	7.0		1044	6.3
	1801	0.7		1807	1.3
	O 2245	6.8		● 2302	6.3
14 TU	0617	0.8	29 W	0625	1.3
	1104	7.0		1115	6.3
	1849	0.7		1842	1.2
	2333	6.8		2333	6.3
15 W	0703	0.8	30 TH	0700	1.3
	1150	6.9		1146	6.3
	1935	0.8		1916	1.3

DECEMBER

	Time	M		Time	M
1 F	0007	6.3	16 SA	0059	6.6
	0735	1.3		0827	1.0
	1218	6.2		1320	6.3
	1949	1.3		2053	1.2
2 SA	0042	6.3	17 SU	0142	6.4
	0812	1.4		0911	1.1
	1253	6.1		1405	6.1
	2026	1.4		2134	1.5
3 SU	0120	6.2	18 M	0227	6.2
	0850	1.4		0955	1.3
	1334	5.9		1453	5.8
	2105	1.6		2214	1.7
4 M	0204	6.0	19 TU	0315	6.0
	0932	1.6		1038	1.6
	1423	5.7		1545	5.6
	2150	1.7		2255	2.0
5 TU	0254	5.9	20 W	0407	5.8
	1020	1.7		1126	1.8
	1525	5.6		1645	5.4
	2242	1.9		2343	2.2
6 W	0356	5.8	21 TH	0508	5.6
	1118	1.7		1219	2.0
	1638	5.5		1753	5.2
	2346	2.0			
7 TH	0506	5.8	22 F	0041	2.3
	1227	1.7		0615	5.4
	1751	5.6		1320	2.1
				1859	5.2
8 F	0059	1.9	23 SA	0147	2.3
	0617	5.9		0721	5.4
	1340	1.6		1426	2.0
	1857	5.7		1959	5.3
9 SA	0208	1.7	24 SU	0256	2.2
	0720	6.1		0818	5.5
	1446	1.4		1527	1.9
	1957	5.9		2050	5.5
10 SU	0311	1.5	25 M	0356	2.0
	0816	6.3		0905	5.7
	1548	1.2		1620	1.7
	2051	6.2		2132	5.7
11 M	0410	1.3	26 TU	0445	1.7
	0910	6.5		0946	5.8
	1645	1.1		1705	1.5
	2143	6.4		2210	5.9
12 TU	0508	1.1	27 W	0530	1.5
	1003	6.6		1024	6.0
	1742	1.0		1747	1.4
	O 2235	6.5		2245	6.1
13 W	0601	1.0	28 TH	0611	1.3
	1054	6.7		1101	6.1
	1835	0.9		1828	1.3
	2326	6.6		● 2322	6.3
14 TH	0653	0.9	29 F	0650	1.2
	1144	6.7		1137	6.2
	1924	0.9		1904	1.2
15 F	0014	6.6	30 SA	0000	6.4
	0741	0.9		0727	1.1
	1234	6.5		1214	6.2
	2011	1.1		1940	1.2
			31 SU	0038	6.4
				0804	1.1
				1253	6.2
				2015	1.2

HEIGHT IN METRES

CORRECTION SERVICE

Up-to-the-minute correction sheets are available free on completion
and
return of one of the forms below (with a large SAE) to:

The Editor: Pilot Packs
Adlard Coles
8 Grafton Street
London W1X 3LA

When a new edition of a Pilot Pack appears, corrections will no longer be available to the previous edition. Corrections supplied after 1 November will include a copy of the Dover Tide Tables for the following year.

Any suggestions or corrections will be most welcome.

Please supply the current correction sheet for the Adlard Coles Pilot Pack Vol 3, 1989 edition

NAME

ADDRESS

SUGGESTIONS OR CORRECTIONS

If you are interested in checking pilotage and port details in your area and reporting changes to the publishers on a regular basis (see the Introduction)

Tick Area proposed

Please supply the current correction sheet for the Adlard Coles Pilot Pack Vol 3, 1989 edition

NAME

ADDRESS

SUGGESTIONS OR CORRECTIONS

If you are interested in receiving details of Pilot Pack Vol 1 (Great Yarmouth to Littlehampton, Ijmuiden to Carentan) and Pilot Pack Vol 2 (Chichester to Portland, The Channel Islands, St Vaast to Erquy) or any other Adlard Coles titles, please complete this form.

I would like details of: Pilot Pack Vol 1 Pilot Pack Vol 2

Other Adlard Coles titles

Name

Address

PLASTIC WALLET FOR YOUR PILOT PACK

A heavy duty, zip closure, transparent plastic wallet is available for this book to enable it to be used in all weathers in the cockpit or on the flying bridge. When opened at the desired page this book may be folded back and inserted in the wallet for viewing on either side. The wallet may also be used for charts in the same way.
Available at £2.95 each, post free from Kelvin Hughes Ltd, 145 Minories, London EC31NH. Tel: 01-709-9076. Enclose cheque made payable to Kelvin Hughes Ltd or give details (card no. and expiry date) of your Access/Eurocard/ Mastercharge/Barclaycard/Visa/American Express/Diners Club Card.